Student Study Guide
to accompany

An Introduction to the Biology of

Marine Life

Sixth Edition

James L. Sumich

Prepared by
Larry M. Lewis
Salem State College

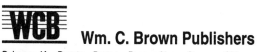

Wm. C. Brown Publishers

Dubuque, IA   Bogota   Boston   Buenos Aires   Caracas   Chicago
Guilford, CT   London   Madrid   Mexico City   Sydney   Toronto

A Times Mirror Company

ISBN 0-697-27977-4

Printed in the United States of America by Times Mirror Higher Education Group, Inc., 2460 Kerper Boulevard, Dubuque, Iowa, 52001

10 9 8 7 6 5 4 3 2 1

Contents

The Philosophy and Use of this Guide   *iv*

1   The Ocean as a Habitat   *1*

2   Some Ecological and Biological Concepts   *8*

3   Marine Phytoplankton   *14*

4   Marine Plants   *20*

5   Primary Production in the Sea   *25*

6   Protozoans and Invertebrates   *32*

7   Marine Vertebrates   *38*

8   The Intertidal   *43*

9   Estuaries   *49*

10   Coral Reefs   *54*

11   Below the Tides   *60*

12   The Zooplankton   *65*

13   Nekton - Distribution, Locomotion, and Feeding   *70*

14   Nekton - Migration, Sensory Reception, and Reproduction   *77*

15   Food from the Sea   *83*

16   Ocean Pollution   *89*

Answer Key   *94*
About the Author   *98*

The Philosophy and Use of this Guide

Introduction

Science education in America is in a state of crisis. While perhaps 50% of our country's future economy is in the areas of science, technology, health, and engineering, only about 15% of college freshman elect to major in one of these subjects. So many students fail to appreciate science that we may not remain competitive as a nation. The 85% of freshman who do not major in science often have a very negative attitude about it. The teaching methods of the past do not seem to be succeeding with many of today's students. The problems of science education in America are complex, and perhaps part of the difficulty is because many college and upper-level high-school students do not arrive in class prepared. I have taught at the college level since 1975, and for some time I have felt a need to provide students with new tools and methods. Frankly, at a personal level, I feel that I survived the educational system of the past and was not as nurtured as I could have been. This guide is the result of much searching. I hope that it helps you in your exploration of the marine world and in other courses as well.

First the Good News

Almost all of you will begin this course in marine biology with great expectations. Perhaps the sea is in your blood, and you look forward to this class with eager anticipation. Our goal is to use that interest and motivation to ensure a successful and enjoyable learning experience.

In addition, your instructor has selected a really good textbook for you to read. I had enjoyed Dr. Sumich's book in my own classroom for some time before I began this study guide. It is an interesting and well-written text, and if you read it before your instructor covers the chapter at issue you will be prepared for class.

Now the Bad News—A Prescription for Failure

Despite the best of intentions and instructors, many of you will do poorly in this course. Approaches to study developed in high school will often fail. Here is my prescription for failure: read the chapter in the textbook before class, come to each class on time and take notes, and "study" before exams. This may sound like a guide to success, but it will not work for many of you. Marine biology, like any science, is quite a challenge. There is a large vocabulary to learn, and many of the words will seem long and strange. Frequently, the organisms will be unfamiliar; alas, more than whales and sharks live in the sea. Many of the creatures we will study are very small and beyond our ordinary experiences. New concepts need to be understood. The chemistry and physics of seawater are not easy to learn. The physiology of an organism adapting to a specific salinity, temperature, and pressure is not trivial. What is a student to do?

Help with the First Challenge: Note Taking

I recently attended a seminar by a chemistry professor who, like myself, was trying to focus on the needs of your generation. He asked the following question: If students come to our classes less prepared than previous generations, does that mean that they have to learn less? His answer was no. If, for example, students are not good at taking notes, his answer was to provide students with good notes. The fact is professional schools often provide students with notes. It is an interesting observation that we commonly provide our best and brightest students in medical schools with a set of class notes but do not provide a similar tool for college freshman. The notion that students should spend their time in class taking notes comes from the Middle Ages when only the instructors had books. More recently, a common belief has developed that the act of taking notes enhances learning. Is this always true? Clearly, a student with a poorly constructed set of notes is not in a good position to study. This book provides you with outlines of the entire book. They really are my lecture notes. While every instructor may not follow them exactly, many or most lectures will have a similar pattern. Bring them to class and add what is missing. Try to spend your time in class actively listening to your instructor. Try to think of questions. Each student should be a thinker, not a secretary. How many times in class have you desperately written a sentence in your book while the instructor was off on some other topic?

Help with the Biggest Challenge: Learning Marine Biology

Learning is an active process; it is something that **you** do—not your instructor. The writing exercises in this book are applications of ideas developed at the Institute for Writing and Thinking at Bard College. Think of this book as a directed journal. After each chapter, you will write a summary. If you do it within 24 hours after your class, you will enhance your retention of the lecture material. This informal writing may be new for you; it is writing to learn. While you write your summary, try to keep your pen or computer active. Do not worry about spelling or grammar. Fill the page; the more you write, the more you will learn.

Second, you will define key words or terms in your own words. Yes, much of this will be a repetition of material in the summary. However, repetition of this nature will reinforce learning. It is important to use your own words and resist the temptation to copy a definition out of your textbook.

Third, you will develop questions. Asking questions is one of the real jobs of the scientist, and in this section you can work on some additional skills of critical thinking. After the previous exercises, questions should arise—you should be able to distinguish between what you understand from what is unclear. Your questions should be real ones. They may form the basis of a visit to your instructor during an office hour. Questions, at times, may transcend the lecture or class in question. Try to make connections between marine biology and other subjects. If no questions come to mind, that may be a sign of a problem. If that happens, write multiple-choice questions similar to those at the end of each chapter in this guide. At the least, when you write a multiple-choice question, you have to decide what subjects are the most important, and what are correct and incorrect answers.

Fourth, write a reflective paragraph at the end of each chapter. As an instructor, I enjoy reading these sections the most. In the technical jargon of educators, we often call this section *metacognition,* or *thinking about thinking.* Here you can reflect upon your learning process. There is evidence that if you can understand how you learn, you will develop into a better student. However, this may not be easy. Try to identify what it was like to learn the material in the chapter. Try to make connections with other subjects. Most importantly, as difficulties develop in a chapter or related lecture, develop a plan to learn the material. The time to plan for difficulties is within 24 hours of the class, not the night before the exam. Here you can develop your skills as a self-advocate. Perhaps you need to reread a section of the textbook. Will a discussion with a classmate help? Consider how lonely instructors often feel during office hours.

In addition, I have included sample multiple-choice and completion questions to test your knowledge. You may wish to answer these questions before exams, as some instructors will use questions of this type. If you have prepared all your journal assignments on a regular basis, there will be no need to memorize the night before the exam. A quick reading of your journals and looking at sample questions should be enough.

If you do all these exercises for each chapter, you will learn marine biology and become a better student. Good luck, and if this little book proves useful, please drop me a line.

Dr. Larry Lewis
Department of Biology
Salem State College
352 Lafayette Street
Salem, Massachusetts 01970

Technology Support: Should you have access to the Internet, a marine resource is available titled "Oceans Online." This listserve has been established to service professionals in oceanography, marine biology, geophysics, geology, engineering, manufacturing, navy, coast guard, civil maritime, maritime law and insurance, and academic interests. For an automated reply of topic listings, send an e-mail to OCEANS@VBS.COM and type INDEX on the subject line.

Chapter Outline

1.0. The Changing Marine Environment
    1.1.   Solar system, 5 billion years old
    1.2.   Earth: aggregation and radiation $==>$ heat
    1.3.   Volcanoes vented liquid and gases of primitive atmosphere
           *   Heavy metals sunk to core
           *   Lighter materials cooled to form crust
           *   Cooling water vapors $==>$ primitive sea
    1.4.   Green plants - oxygen - by 600 million years ago, 0.2% O_2
           *   New forms of metabolism possible
           *   By 500 million years ago, worms, sponges, corals, etc.
           *   No land life because of UV radiation
    1.5.   As O_2 increased, some converted to O_3 = ozone. By 400 million years ago, land life possible
    1.6.   At first, Pangaea - seafloor spreading - plate tectonics
           *   Rigid plates float on liquid mantle
           *   1977, Alvin; hot-water vents part of oceanic ridges
           *   Now, South Atlantic widening 5–6 feet per lifetime
           *   Pacific shrinking faster
    1.7.   Most fossils destroyed by conveyor belt of ocean floor
           *   Most marine fossils found on land that was ancient seabed
2.0. The World Ocean
    2.1.   70% earth's surface
    2.2.   Average depth, 3,800 m
    2.3.   One interconnected ocean system
    2.4.   Northern Hemisphere has two-thirds of land area
    2.5.   Southern Hemisphere = 80% water
    2.6.   Continental shelf with shelf break at edge = (120-200 m depths)
    2.7.   Continental slope (3,000-4,000 m depths)
    2.8.   Abyssal plains + oceanic ridge and rise systems = 30% ocean basin
    2.9.   Trenches, 6,000$^+$ meters deep. Mariannas, 11,022 m
    2.10.   Seamounts - oceanic islands - volcanic mountains
           *   Capped by coral atolls or fringed by coral reefs
3.0. Properties of Seawater
    3.1.   80–90% of marine organisms
    3.2.   Provides buoyancy
    3.3.   Medium for chemical reactions of life
    3.4.   Pure water
           *   Forms hydrogen bonds
    3.5.   Viscosity and surface tension
    3.6.   Density-temperature relationships
           *   At 4° C or above, density increases with decreasing temperature
           *   Below 4° C to 0° C reversed; 8% less dense as ice
    3.7.   Heat capacity
           *   Heat measured in calories

3.8.   Solvent action
*   Breaks ionic bonds, salts dissolve
3.9.   Seawater
*   96.5% pure water, 3.5% dissolved compounds
*   Inorganics: salts
*   Dissolved gases
*   Organics: fats, carbohydrates, vitamins, etc.
*   Synthetics: DDT, PCBs
3.10.   Dissolved salts
*   Salinity, parts per thousand = 0/00
*   Average = 350/00
*   Near 0 at mouth of rivers to 400/00 in Red Sea
*   Affected by evaporation, precipitation, rivers, freezing
*   Salts exist as ions
3.11.   Light and temperature in the sea
*   Light energy used for vision and photosynthesis; visible light
*   Amount of light energy reaching sea depends upon atmospheric conditions and angle of sun
*   Photic zone supports photosynthesis
*   Depth depends upon dissolved substances, sediments, plankton
*   Sunlight absorbed is converted to heat energy, increases molecular activity
*   High heat capacity limits marine temperature to a narrow range compared with the land
*   Deep sea vents, some 60° C and above
*   Mostly 0-30° C
*   Temperature affects distribution of life-forms
3.12.   Salinity-temperature-density relationships
*   Dense, cold bottom water
*   Warmer, well-mixed surface water
*   Thermocline between limits exchange of gases and nutrients
3.13.   Pressure
*   Increases 1 atm per 10 m in depth
*   1,000 atm in trenches
3.14.   Dissolved gases and acid/base buffering
*   More soluble at lower temperatures
*   Great capacity to absorb CO_2 because it combines with water to form carbonic acid
*   pH reflects H^+ ions
*   Buffered by carbonic acid-bicarbonate-carbonate system
* O_2, green plants near surface
*   By 1,000 m, O_2 minimum zone
3.15.   Dissolved nutrients
*   Nitrate and phosphate = fertilizers of sea
*   Used first by photosynthetic organisms
*   Vertical distribution inverse of oxygen concentration—why?

4.0. The Ocean in Motion—Waves
4.1.   Differential surface heating = = > winds
4.2.   Winds = = = > waves (3 mm to 30^+ m)
4.3.   Height, length, period
4.4.   Energy transmitted forward, particles do not move
*   In shallow areas, bottom friction causes waves to become higher, steeper, pitch forward, and break—shapes shoreline

5.0. Tides
5.1.   Long period waves
5.2.   Periodic rise and fall of sea surface against coastlines

5.3.   Gravity of moon exerts 2X force as sun
5.4.   Often semidiurnal but complex patterns arise because of regional interactions with land
5.5.   Surface currents
    *   Reflects major wind belts, up to 200 m deep
    *   Water flow not parallel to wind; deflected to right in
    *   Northern Hemisphere; Coriolis Effect
5.6.   Vertical water movements
    *   Sinking, colder latitudes
    *   Circulates oxygen, slow—100s—1,000s of years
    *   Upwelling brings nutrient-rich water to surface
6.0. Classification of the Marine Environment
6.1.   Photic zone: photosynthesis, 50–100 m
6.2.   Aphotic zone
6.3.   Benthic division
6.4.   Pelagic division
6.5.   Neritic province
6.6.   Oceanic province

Journal Questions

1. Summary

Write a summary of the chapter. Remember that a ''Summary'' is not just a statement of the topics in the chapter or lecture. The goal is to review the key points as an extensive narrative. Let your pencil or the keys of your computer fly and include a description of the origin of the ocean. Then consider the major regions of the ocean. Note the difference between the Northern and Southern Hemispheres. Consider the properties of seawater, including how salinity is measured and the average salinity of the ocean. What types of motion do we find in the sea? What causes these movements, and how are they related to global temperature and the shaping of coastlines?

2. Define the following key words in your own words. Please feel free to add additional words to this list.

a. photic zone
b. benthic division
c. pelagic division
d. continental shelf
e. continental drift

f. abyssal plains
g. oceanic ridge and rise systems
h. hydrogen bond (between water molecules)
i. thermocline
j. spring tide

3. Question Section

Try to write meaningful questions. Do you understand why what we call the continental shelf today was once dry land? Do you know what heat capacity is, and the role that the oceans play in regulating global temperature? Do you understand enough chemistry to follow the chapter? Can you explain the color of the sea?

4. Reflections

What is it like to learn the material in this chapter? How is it similar or different from the other classes you are taking? If some of the material is difficult for you to understand, indicate the sections and try to write why you are having difficulty. Try to develop a plan to learn what is a problem for you. Perhaps you need to reread your textbook, or perhaps it may be valuable to visit your professor during an office hour. You may also wish to write about how the class makes you feel or any connections to other classes.

Practice Questions

Multiple-Choice Questions (see back of book for answers)

1. The solar system is approximately _____ years old.
   a. 1 million
   b. 5 million
   c. 1 billion
   d. 5 billion
   e. 1 trillion

2. Marine life greatly increased in diversity between 600 and 500 million years ago because
   a. it was during that time that the salt content in the sea first became high enough to support marine life.
   b. CO_2 levels were too great before 600 million years ago.
   c. the O_2 level in the atmosphere was finally high enough to support large aerobic organisms.
   d. the temperature of the earth was finally cool enough to support life.
   e. it was during this time that condensation of water vapor first produced the ocean.

3. Why are relatively few marine fossils found?
   a. Many are destroyed as regions of the ocean floor sink into trenches and are melted by the heat of the earth's core.
   b. Few people try to find marine fossils.
   c. Marine fossils are buried by thick layers of sediment.
   d. The high-salt content of the sea prevents dead organisms from forming fossils.
   e. The cold temperature of the deep-sea floor prevents fossils from forming.

4. The first global explorer to know his exact latitude and longitude at sea was
   a. Pytheas.
   b. Eratosthenes of Alexandria.
   c. James Cook.
   d. James Watson.
   e. Jacques-Yves Cousteau.

5. The Pacific Ocean contains nearly ____ of the total ocean area.
   a. 10%
   b. 25%
   c. 50%
   d. 75%
   e. 90%

6. Hydrogen bonding between adjacent water molecules results in
   a. a viscosity that permits organisms to sink very quickly.
   b. a thick molecular skin that increases in thickness as water temperature increases.
   c. a density that increases as the water temperature decreases—it is most dense at 0° C.
   d. all of the above (read each of the above carefully).
   e. none of the above (you can't fool me!).

7. When seawater freezes, the salinity of the surrounding water
   a. decreases because much salt becomes incorporated in the sea ice, thus, the sea ice floats.
   b. decreases because it is too cold to rain when sea ice forms.
   c. increases because it is too cold to rain when sea ice forms.
   d. increases because salt is excluded from the ice crystals.
   e. remains the same because the concentration of salt in sea ice is the same as in sea water.

8. Which wavelengths of visible light penetrate most deeply into the sea?
   a. red
   b. blue and green
   c. orange
   d. yellow
   e. pink

9. New oceanic crust originates at
   a. the axes of oceanic ridge or rise systems.
   b. the axes of oceanic trenches.
   c. island arc systems.
   d. large terrestrial mountain ranges.
   e. upwellings.

10. The Coriolis effect results from
    a. earth's gravitational field.
    b. earth's magnetic field.
    c. lilies of the field.
    d. rotation of the earth.
    e. all of the above.

11. Large bodies of water have a moderating effect on the climate of nearby areas because of the water's
    a. solvent properties.
    b. high heat capacity.
    c. transparency.
    d. wetness.
    e. motion.

12. The average salinity of open-ocean seawater is about
    a. 3.5 %.
    b. 35%.
    c. 3.0%.
    d. 30%.
    e. 85 proof . . . burp!

13. _____ are probable areas where excess oceanic crust is absorbed back into the mantle.
    a. The axes of oceanic ridge or rise systems
    b. The axes of oceanic trenches
    c. Island arc systems
    d. Large mountain ranges
    e. Typhoon lagoons

14. What causes a thermocline to disappear in temperate regions?
    a. when cold winter temperatures cool the surface water to the same temperature as deeper water
    b. when fall winds cause rapid motion of surface waters, which causes the thermocline to break down
    c. when the strong summer sun causes a warming that breaks the thermocline down
    d. when spring storms cause surface waters to move deep into the sea
    e. all of the above, it depends upon local conditions

15. Why is carbon dioxide generally abundant in the sea?
    a. because it reacts with the salt in seawater
    b. because it combines with water to form a weak acid—carbonic acid
    c. because the acid nature of seawater easily combines with basic CO_2.
    d. all of the above
    e. none of the above

Complete the sentences below by writing the correct answer in the blank space (see back of book for answers).

1. Water over the continental shelves is said to be the _____ province.
2. Region of the sea where sunlight does not penetrate is the _____ zone.
3. Deep nutrient-rich waters can be brought to the surface by _____ processes.
4. When a continental land mass obstructs the flow of a current, such that the water flows directly back against the established current, it is said to be a _____.
5. When the moon and the sun are at right angles to each other, a moderate tide results, known as a _____ tide.
6. The most important fertilizers of the sea are the nutrients _____ and _____.
7. Waves and surface currents are caused by _____.
8. The carbonic acid-bicarbonate system in seawater serves to limit changes in pH and serves as a _____.
9. The boundary between the continental mass and the true ocean floor is the _____ _____.
10. At first, life evolved in the sea when organisms could be protected from the harm of _____ radiation.

Chapter Outline

1.0. Introduction: Four Rs of Living
    1.1.   Respiration
    1.2.   Reproduction
    1.3.   Response
    1.4.   Regulation
2.0. The Cellular Structure of Life
    2.1.   Procaryotes
        *   Bacteria and Cyanobacteria
           **   Cell wall
           **   Plasma membrane
           **   Cytoplasm
           **   Flagellum
           **   Single chromosome
    2.2.   Eucaryotes
        *   Nucleus = nuclear envelope + chromosomes
        *   Mitochondria
        *   Endoplasmic reticulum
        *   Chloroplasts
3.0. Adaptations of Marine Life
    3.1.   Introduction
        *   Ecological adaptations: by individuals, immediate results; organism secures sufficient resources and reproduces
        *   Evolutionary adaptations: by populations over many generations
        *   All populations reproduce at a greater rate than resources can support; those better equipped survive and transmit genes to next generation
        *   Winners in the sea must adapt to changes in temperature, salinity, available oxygen, light, food, and neighbors' efforts
    3.2.   The value of sex
        *   Transfers genes through the generations
        *   Asexual reproduction
        *   Sexual reproduction results in diverse offspring
        *   Diverse offspring, less competition than identicals; advantage when population exceeds one million.
        *   Sexual reproduction, complex and costly
        *   Hermaphroditic organisms (i.e., barnacles and some fish) may be simultaneous or sequential
        *   Meiosis results in haploid gametes
        *   Fertilization results in diploid zygotes
    3.3.   Salinity effects
        *   Homeostasis
        *   Body fluids regulated by selectively permeable membranes
        *   Diffusion moves water and other molecules across membranes
        *   Sea cucumber has internal fluids 350/00 = seawater = isotonic

* Now place sea cucumber in lake, body fluids now hyperosmotic. What will be the net movement by osmosis?
* Where can these animals live?
* Other organisms (i.e., bony fish and plants), osmoregulatory mechanisms, require energy
* Salmon's body fluids, 180/00.
  ** In salt water, salmon drink a lot and secrete excess salts by chloride cells in gills
  ** In fresh water, they drink little and kidneys produce lots of dilute urine. Needed salts actively absorbed by cells in gills

3.4. Temperature effects
* Most cannot regulate temperature = poikilotherms or ectotherms. Thus, restricted to narrow optimal temperature ranges
* Metabolic rate doubles every 10° C
* Birds and mammals = homeotherms (40° C) = endotherms
* Large tuna, billfishes, and sharks - intermediate

3.5. Trophic relationships
* ATP, unit of energy exchange
* ATP --> ADP + Pi + energy
* Photosynthesis
* Autotrophs = primary producers = first trophic level. Most use photosynthesis, others chemosynthetic
* Heterotrophs = consumers + decomposers
* Herbivores, second trophic level
* Carnivores, third and higher trophic levels
* Decomposers (bacteria and fungi) feed on detritus
* Biogeochemical cycles link organisms to nonliving reservoirs of nutrients
* Less than 1% of available solar energy absorbed by autotrophs. Less than that available for animals. Why?
* Energy transfer from one trophic level to another = 6–20%
* Food chains follow paths of energy and nutrients
* Food web may be a better term
* Symbiosis
  ** Commensalism
  ** Mutalism
  ** Parasitism

4.0. Spatial Distribution
  4.1. Benthos = epifauna (live on bottom) + infauna (sediment)
  4.2. Nekton = large swimming animals
  4.3. Plankton = phytoplankton + zooplankton

5.0. The General Nature of Marine Life
  5.1. Movement of water mixes food and waste and transports organisms. Even the smallest plankton can become widely distributed
  5.2. Marine biology is mostly the biology of the very small
  5.3. Phytoplankton more dilute than a cornfield = food is limiting ==> most heterotrophs live in photic zone. Why?
  5.4. Many products of primary producers dissolve in water and become food for bacteria. They in turn become food for suspension feeders
  5.5. Yet, giant squid = 15-30 m and blue whales = 200 tons

Journal Questions

1. Summary

Write a narrative based upon the chapter. Include a brief review of cell structure. Carefully reflect upon the adaptations of marine life and consider the value of sex and especially note salinity effects. Based upon the figures in your text, review the osmotic events in the sea cucumber and salmon in fresh water and seawater. Also include a discussion of trophic relationships, the spatial distribution of life in the sea, and end with a discussion of the general nature of marine life.

2. Define the following key words in your own words. Please feel free to add additional words to this list.

a. plasma membrane
b. procaryote
c. eucaryote
d. simultaneous vs. sequential hermaphrodite
e. homeostasis
f. osmosis
g. hypoosmotic
h. poikilotherm

i. primary producer
j. heterotroph
k. biogeochemical cycle
l. symbiosis
m. commensalism
n. epifauna
o. plankton
p. nekton
q. benthos

3. Question Section

Try to think of meaningful questions. Do you understand the difference between a procaryotic and eucaryotic cell? Is it clear to you how a sea cucumber and a salmon face the problem of osmoregulation? Can you appreciate why the marine food web is different from the trophic relationships we find in land-based ecosystems? Are there points in your lecture that you found unclear?

4. Reflections

What is it like to learn the material in this chapter? Compare learning about cell structure to learning about the value of sex in the sea to understanding the effects of salinity. What are the greatest challenges presented by this chapter? What plan can you develop to feel sure that you can learn all the information well? Feel free to add any other comments or reflections about your classes or text concerning this chapter.

Practice Questions

Multiple Choice Questions (see back of book for answers)

1. What attribute of life includes enzyme-controlled reactions and results in the useful energy-rich molecules necessary for life?
   a. respiration
   b. reproduction
   c. response
   d. regulation
   e. all of the above

2. What part of a cell regulates the exchange of material between the cytoplasm and external environment?
   a. cell wall
   b. flagellum
   c. nuclear membrane
   d. endoplasmic reticulum
   e. plasma membrane

3. In the sea, large and diverse life-forms are said to be
   a. eucaryotes.
   b. procaryotes.
   c. mitochondria.
   d. all of the above.
   e. none of the above.

4. What is the most important point of sexual reproduction?
   a. to produce identical copies of the parents
   b. to produce large numbers of offspring
   c. to produce diverse offspring
   d. to reproduce with the least cost of biological energy
   e. to ensure the survival of small populations

5. If a sea cucumber with internal fluids of 350/00 is placed into a lake, the body fluids of the sea cucumber are now _____ with respect to the lake.
   a. isomotic
   b. hypoosmotic
   c. hypodermic
   d. hyperosmotic
   e. hypnotic

6. Sea animals that have a nearly constant body temperature are said to be
   a. homeotherms.
   b. ectotherms.
   c. ectoplasms.
   d. poikilotherms.
   e. all of the above.

7. The fundamental molecule of energy exchange in all cells is
   a. ADP.
   b. ATP.
   c. STP.
   d. chlorophyll.
   e. sugar.

8. At the first trophic level one may find
   a. primary producers.
   b. chemosynthetic autotrophs.
   c. photosynthetic organisms.
   d. all of the above.
   e. none of the above.

9. Autotrophic plankton are said to be
   a. nekton.
   b. zooplankton.
   c. phytoplankton.
   d. all of the above.
   e. none of the above.

10. Atmospheric greenhouse gases include
    a. water vapor.
    b. CO_2.
    c. methane.
    d. all of the above.
    e. none of the above.

Complete the sentences below by writing the correct answer in the blank space (see back of book for answers.

1. In many cells, photosynthesis occurs in structures termed _____.
2. Adaptations of an organism to the environment during the lifetime of the organism are said to be _____ adaptations.
3. Organisms that reproduce by making offspring identical to themselves are said to engage in _____ reproduction.
4. When a sea cucumber is placed in water where the concentration of salt in the external fluid is equal to the internal salt concentration, the condition is known as _____.
5. When a salmon swims in seawater, chloride cells in the gills will _____ salts.
6. The movement of a substance from a region of high concentration to a region of low concentration is called _____.
7. Animals in the sea that lack mechanisms to regulate body temperatures are said to be _____.
8. Organisms that live in marine environments that lack oxygen are found to engage in _____ respiration.
9. Decomposers feed mainly upon _____.
10. Organisms that feed upon minute food particles are said to be _____ _____.

Chapter Outline

1.0. The Five Kingdoms and Taxonomic Classification; Margulis, 1978
    1.1.   Kingdom Monera = procaryotes = Bacteria + Cyanobacteria
    1.2.   Kingdom Protista = eucaryotic single-celled or aggregates
    1.3.   Kingdom Fungi = hyphae with cell walls of chitin
    1.4.   Kingdoms Plantae and Animalia = multicellular
    1.5.   1.5 million species identified; 10–30 million species exist
    1.6.   Most species land = living insects; marine still diverse
    1.7.   Taxonomy, reflects evolutionary or phylogenic relationships
    1.8.   Each group = taxon (taxa pl.)
    1.9.   Basic unit = species
    1.10.   Linneaus, genus species, i.e., *Delphinus delphis*
    1.11.   Kingdom, Phylum(Division), Class, Order, Family, Genus, Species
2.0. Phytoplankton
    2.1.   Marine primary producers = Cyanobacteria --> kelps
    2.2.   Marine phytoplankton = kingdoms Monera, Protista, and Plantae
    2.3.   Divisions: photosynthetic pigments, cell-wall structures, storage products, and growth forms
    2.4.   Single-celled microscopic organisms in photic zone
    2.5.   Nannoplankton (5–20 μm) most important
3.0. Cyanobacteria
    3.1.   Small, less than 5 μm
    3.2.   Procaryotes with chlorophyll *a* similar to eucaryotes
    3.3.   Intertidal and estuarine areas
    3.4.   Dense blooms in warm waters
       *   Red phycobilin, Red Sea
       *   Fossil stromatolites, over 3 billion years old
    3.5.   Benthic Cyanobacteria
       *   May form macroscopic colonies
       *   Tarlike patches on rocks in intertidal zone
       *   Mudflats of coastal estuaries
       *   Associations with coral reefs
       *   Nitrogen fixation
       *   Symbionts and epiphytes
4.0. Chrysophyta
    4.1.   Single-celled, flagellated, chlorophylls *a* and *c* and xanthophylls
    4.2.   Cell walls or skeletons of silica or calcium carbonate
    4.3.   Class Chrysophyceae, nannoplankton, mostly fresh water
       *   Coccolithophores
          **   Small calcareous plates = coccoliths
          **   Warm and temperate seas
       *   Silicoflagellates, internal ornate skeletons
    4.4.   Class Bacillariophyceae, diatoms
       *   Often most abundant phytoplankton
       *   Unicellular (15μm–1 mm), chains, or loose aggregates
       *   Cell wall = frustule = epitheca + hypotheca

          *   Frustule up to 95% silica + some pectin
          *   Centric and pennate forms
          *   Each pit in cell wall = areolus
          *   Cell division reduces cell size until an auxospore forms

5.0. Dinophyta

    5.1.   Unicellular (25–1,000 μm), two flagella, photosynthetic pigments similar to diatoms

    5.2.   Wall made of articulating cellulose plates

    5.3.   Bioluminescence

    5.4.   Cause blooms, red tides, paralytic shellfish poisoning (saxitoxin)

6.0. Special Adaptations for a Planktonic Existence

    6.1.   Need to obtain nutrients and light

    6.2.   Cells dispersed to utilize nutrients, yet restricted to photic zone

    6.3.   Size
          *   Small size; must get nutrients and deposit waste in same environment by diffusion
          *   Surface/volume ratios enhance diffusion
          *   Favors small size or complex shapes
          *   Cell vacuoles common

    6.4.   Sinking
          *   Heavy cell walls more dense than seawater.
          *   Need to sink slowly
              **   Small size, horns, wings, etc.
              **   Complex shapes may cause zigzag or spiral sinking patterns
              **   Planktonic Cyanobacteria - gas vacuoles

    6.5.   Adjustments to unfavorable environmental conditions
          *   Horns, spines, and slime discourage herbivores
          *   Flagella help a cell move from its own waste
          *   Toxins of dinophytes may discourage predation
          *   Diatoms form cysts

Journal Questions

1. Summary

Write a narrative based upon the chapter. Review the five kingdoms of life on earth and consider which kingdoms include the phytoplankton. Briefly review each of the phyla or divisions that comprise the phytoplankton communities of the sea. Most importantly, review the special problems faced by phytoplankton and how they have adapted to solve these difficulties.

2. **Define the following key words in your own words.** Please feel free to add additional words to this list.

a. Monera
b. Protista
c. taxa
d. picoplankton

e. epiphytes
f. coccoliths
g. frustule

3. Question Section

Create questions that consider the details of taxonomy as well as the adaptations of phytoplankton. You may wish to write a question or two concerning how different types of microscopes can image phytoplankton.

4. Reflections

While large marine forms draw much of our interest to the sea, it is upon the tiny phytoplankton that all life in the sea depends. How has the subject matter in this chapter changed your thinking about the sea? Is it possible for you to relate to the problems of sinking plankton? What special difficulties exist in trying to appreciate the very small?

Practice Questions

Multiple-Choice Questions (see back of book for answers)

Match the type of organism to its correct kingdom.
1. Body composed of threadlike tubes call hyphae
2. Eucaryotic, single-celled, and photosynthetic
3. Procaryotic, single-celled, and photosynthetic

    a. Plantae
    b. Animalia
    c. Fungi
    d. Protista
    e. Monera

4. The correct hierarchical sequence of taxonomic categories is
    a. class, order, genus, family, species.
    b. order, class, species, family, order.
    c. species, family, order, genus, class.
    d. class, order, family, genus, species.
    e. order, family, class, genus, species.

5. What type of microscope requires that specimens be very thin?
    a. transmission electron microscope
    b. scanning electron microscope
    c. dissecting microscope
    d. reflected light microscope
    e. all of the above

6. What division or phylum of phytoplankton is important for nitrogen fixation?
    a. Dinophyta
    b. Chrysophyta
    c. Cyanobacteria
    d. Chlorophyta
    e. Ciliophora

7. What type of phytoplankton have cell walls composed mostly of silica, are unicellular, and are often the most abundant class of phytoplankton found?
    a. coccolithophores
    b. silicoflagellates
    c. diatoms
    d. dinophytes
    e. cyanobacteria

8. Members of the Dinophyta
    a. may show bioluminescence.
    b. can cause blooms.
    c. may be covered by cellulose plates with spines or horns.
    d. all of the above.
    e. none of the above.

9. How can phytoplankton reduce their rate of sinking?
    a. by growing to a larger size
    b. by growing a smooth and perfectly round cell wall
    c. by never producing internal gases
    d. all of the above
    e. none of the above

Complete the sentences below by writing the correct answer in the blank space (see back of book for answers).

1. Electron microscopes can achieve greater resolution than light microscopes because they use radiation with

_____ _____.

2. In addition to chlorophyll *a* and carotenes, cyanobacteria contain photosynthetic pigments known as

_____.

3. Phytoplankton with small calcareous plates, or coccoliths, belong in the division _____.

4. The cell wall of a diatom consists of two halves, the larger _____ and the _____.
5. Phytoplankton with two flagella, where one is ribbonlike and is found in a transverse groove, belong to the division _____.

Chapter Outline

1.0. Introduction
    1.1.   Macroscopic, multicellular, confined to photic zone
    1.2.   Need solid bottom
    1.3.   Major role at first trophic level within near-shore communities
2.0. The Seaweeds
    2.1.   Macroscopic members of Chlorophyta, Phaeophyta, and Rhodophyta
    2.2.   Abundant in intertidal zones to 30–40 m
    2.3.   Photosynthetic pigments
        *   Division specific
          **   Green algae: simple filaments, flat sheets, to branching forms (i.e., *Codium,* up to 8 m)
          **   Brown algae: xanthophylls, fucoxanthin
          **   Red algae: phycobilins
    2.4.   Structural features of seaweeds
        *   Lack roots, flowers, seeds, and true leaves
        *   May have blade, stipe, and holdfast
        *   The blade
          **   Unlike higher plants, cells on surface are more active in photosynthesis than deeper ones
          **   Each surface similar in structure and function
        *   Pneumatocysts—gas-filled floats
          **   *Sargassum* in Sargasso sea
          ** N_2, O_2, CO_2, and CO
        *   The stipe
          **   May transport photosynthetic products down plant
        *   The holdfast
          **   Short rootlike haptera
          **   *Penicillus,* fine hairlike haptera
    2.5.   Reproduction and growth
        *   Sexual or asexual
        *   Alternation of sporophyte and gametophyte
          **   Identical generations in *Ulva*
          **   In *Codium* and *Fucus,* multicellular gametophyte missing
          **   *Laminaria,* sporophyte dominates
        *   Red algae: no flagellated reproductive cells, 3 generations (includes unique carposporophyte)
        *   Development by mitosis at meristematic regions
        *   *Macrocystis* growth rate, 30 cm per day
3.0. Anthophyta
    3.1.   Flowering plants common near seashores
    3.2.   Sea grasses often submerged
    3.3.   Plants of the salt marsh and mangroves may be emergent
    3.4.   45 species of sea grasses, e.g., *Zostera,* or eel grass

3.5.   Horizontal rhizomes
3.6.   Use water to disperse pollen
3.7.   Geographic distribution
    *   Reds more common in tropics and subtropics
    *   Deposits of $CaCO_3$ more common in warm water; less soluble at high temperature
    *   Browns mostly temperate
    *   Temperate salt marshes replaced by warmer mangal thickets
3.8.   Plant-dominated marine communities
    *   Plant communities rarely dominate marine communities - mangal, salt marshes and sea grasses, and kelp beds
    *   Mangals and salt marshes are emergent plant communities
        **   In mangal, leaves fall into the water and provide energy
        **   In salt marshes, grasses are lush pastures for shellfish and finfish, yet subject to dumping, dredging and filling

Journal Questions

1. Summary

Review the divisions of macroscopic marine plants and consider relative abundance and photosynthetic pigments. Explain the general structure of seaweeds and how their morphology differs from most land plants. Consider the concept of alternation of generations and other aspects of reproduction. How are the red algae unique? Consider variations in the geographical distribution of macroscopic algae and higher plants in the sea, and review the communities where large photosynthetic organisms dominate the landscape.

2. Define the following key words in your own words. Please feel free to add additional words to this list.

a. Phaeophyta

b. kelp

c. blade

d. pneumatocyst

e. haptera

f. gametophyte

g. carpospores

h. mangal

3. Question Section

Write questions that demonstrate your understanding of the variety, morphology, and reproduction of marine seaweeds and higher plants. Include questions that focus on the marine communities dominated by macroscopic photosynthetic organisms.

4. Reflections

Identify why the diversity and reproduction of marine plants may be difficult to learn. What is the best approach to learn the concept of alternation of generations? How can one clearly distinguish all the divisions of marine plant life? How important are these organisms to the marine environment?

Practice Questions

Multiple-Choice Questions (see back of book for answers)

1. A structure that contains a concentration of gases similar to that found in the atmosphere and keeps algal blades in bright light is a
   a. rhizome.
   b. pneumatocyst.
   c. frustule.
   d. frustrated.
   e. slender spine.

2. Meiosis is the reproductive process that produces _____ in species exhibiting alternation of generations.
   a. spores
   b. sporophytes
   c. gametes
   d. zygotes
   e. oldgoats

3. In plants with alternation of generations, the chromosome number of the sporophyte stage is _____ as great as in the gametophyte stage.
   a. just
   b. 0.5 times
   c. 2 times
   d. 4 times
   e. 8 times

Match the division to its correct description.
4. Anthophyta _____
5. Rhodophyta _____
6. Phaeophyta _____

    a. mostly macroscopic marine forms with phycobilins present
    b. macroscopic forms with chlorophylls *a* and *b* and starch present
    c. macroscopic forms with chlorophylls *a* and *c* and laminarin starch present
    d. mostly microscopic marine forms with phycobilins present
    e. mostly microscopic marine forms with chlorophylls *a* and *b* present

7. The most obvious phase in the life cycle of *Laminaria* and other kelps, is the
    a. sporophyte.
    b. gametophyte.
    c. gamete.
    d. zygote.
    e. true root.

8. Shrubby treelike plants that dominate warm coastal marine communities are called
    a. maple trees.
    b. mangroves.
    c. *Spartina.*
    d. calcareous red algae.
    e. *Fucus.*

Complete the sentences below by writing the correct answer in the blank space (see back of book for answers).

1. The part of a marine algae that attaches the plant to the substrate is called a _____.
2. Most marine plants belong to the divisions _____ and _____.
3. The deepest living seaweeds belong to the division _____.
4. The brown color of some seaweeds is due to the class of pigments termed _____.
5. The diploid phase in the life cycle of most seaweeds is called the _____.
6. Plant tissue where cell division is common is _____.
7. Sea grasses grow horizontal structures called _____.
8. In general, red algae are more common in _____ water, while brown algae are more abundant in _____ waters. (Hint: temperature?)

5 **Primary Production in the Sea**

Chapter Outline

1.0. Introduction
  1.1. Benthic, large, long-lived, attached plants produce 5–10% of plant material produced in the sea. Most produced by pelagic phytoplankton
  1.2. Benthic plants have high standing crop = amount of plant material alive at a given point in time
  1.3. Primary production = making organic matter from CO_2 + H_2O + other nutrients transferred to higher trophic levels
  1.4. Gross primary production = total amount of organic material produced by photosynthesis in the sea
  1.5. Net primary production = total amount of organic material available to support consumers
  1.6. Measured in gC/m^2/day or year, where gC = 10% live weight
2.0. Measurements of Primary Production
  2.1. In theory, measure a chemical component of the photosynthetic reaction; i.e., O_2 production or CO_2 consumption.
  2.2. Historically, measure light and dark bottle O_2. Difference = gross primary production
  2.3. O_2 in light bottle is a measure of net primary production
  2.4. But this assumes O_2 consumption equal in both conditions—what about bacteria and zooplankton?
  2.5. More recently, use uptake of C14 as bicarbonate
  2.6. Chlorophyll *a* concentrations = standing crop
  2.7. Relationship between standing crop and productivity depends upon turnover rate
  2.8. Today, primary productivity is measured with coastal zone color scanner aboard weather satellite
3.0. Factors that Affect Primary Production
  3.1. If all parameters for growth unlimited, growth exponential
  3.2. In nature, at least one limiting factor
  3.3. Light
    * Depth of photic zone determined by capacity of sunlight to penetrate seawater
    * Affected by atmospheric conditions (i.e., dust and clouds), angle of sunlight, and water transparency
    * Penetration of light a function of wavelength, reds and violets absorbed first
    * Blue and green (450–550 nm may reach 100 m)
    * Coastal regions green light reduced to 1% in less than 30 m
    * Critical depth = photosynthesis = respiration = to 1% surface intensity
    * Photoinhibition at high light intensities
    * Zone of saturation light intensity = photosynthesis no longer increases with increasing light intensity. Note: some (i.e., dinophytes) better adapted to high light conditions
  3.4. Photosynthetic pigments
    * Chloroplasts contain two separate pigment systems for the light reactions of photosynthesis
    * Pigment systems in grana; stroma contains dark reactions
    * Accessory pigments absorb in areas of the spectrum where chlorophyll cannot
  3.5. Nutrient requirements
    * More complex than the equation below implies
$$6CO_2 + 6H_2O \longrightarrow C_6H_{12}O_6 + 6O_2$$
    * Reflects general needs of cells; i.e., silica, nitrates, and phosphates may limit growth in the photic zone
    * Trace elements: iron, manganese, cobalt, zinc, and copper

　　　　　* 　Primary producers require vitamins
　3.6. 　Nutrient regeneration
　　　　　* 　Biomass of producers returns to environment as fecal, urea, and NH_3 waste. A rain of fecal matter deposed by bacteria as it sinks $==>$ much of the available nutrients wind up beyond the photic zone
　　　　　* 　Thus, primary producers depend upon turbulent mixing (wind and tide) and upwelling to provide nutrients
　　　　　* 　Tropical and subtropical waters have strong thermoclines that inhibit upward mixing $==>$ low rates of primary productivity compared to terrestrial deserts
　　　　　* 　The winter in temperate climates cools surface waters, which then sink, causing the thermocline to disappear. Thus, deep, nutrient-rich waters mix with surface waters. In the spring, the thermocline is reestablished. Hence, convective mixing is a seasonal phenomenon
　　　　　* 　High latitudes promote year-round mixing, but productivity limited by light intensity
　　　　　* 　Coastal upwelling caused by the winds also returns subsurface waters to the photic zone. Four major coastal upwelling areas: California, Peru, Canary, and Benguela Currents
　　　　　* 　Langmuir cells occur between pairs of counter-rotating convention cells. Wind-driven and shallow (3 m), but zooplankton may be 100X more dense.
　3.7. 　Grazing
　　　　　* 　May quickly reduce standing crop
　　　　　* 　Usually, populations are stabilized by feedback mechanisms
　　　　　* 　In nature, dense patches of phytoplankton alternate with zooplankton
　　　　　* 　Some populations of zooplankton seem attracted to one type of phytoplankton and are repulsed by another

4.0. Seasonal Patterns of Marine Primary Production
　4.1. 　Introduction
　　　　　* 　Grazers Nutrient abundance
　　　　　* 　Near-surface turbulence
　　　　　* 　Light intensity - seasonal
　4.2. 　Temperate seas
　　　　　* 　Spring diatom bloom: light increase, abundant nutrients, grazing pressure reduced
　　　　　* 　Begins at poles, diatoms begin to deplete nutrients and grazers begin to reduce diatom populations
　　　　　* 　Dinophytes replace diatoms
　　　　　* 　120 $gC/m^2/yr$—mostly diatoms
　4.3. 　Warm seas
　　　　　* 　A continuous summer
　　　　　* 　Good sunlight, but thermocline limits nutrients
　　　　　* 　40 $gC/m^2/yr$ more dinophytes than diatoms
　4.4. 　Coastal upwelling
　　　　　* 　Can result in massive productivity; i.e., Peru current (if no El Niño) 300 $gC/m^2/yr$
　4.5. 　Polar seas
　　　　　* 　No thermocline, productivity limited by light
　　　　　* 　During summer, huge phytoplankton populations
　　　　　* 　Ice melt in spring releases frozen phytoplankton, which bloom. By early summer, diatoms, krill, birds, seals, fish, and whales.
　　　　　* 　25 $gC/m^2/yr$
　　　　　* 　Major upwellings around Antarctica support most of its populations

5.0. Predictive Modeling
- 5.1. Riley developed equations to predict growth curve for phytoplankton on Georges Bank for a full year
- 5.2. Current models plus or minus 25%

6.0. Global Marine Primary Production
- 6.1. High productivity in summer at high latitudes, shallow regions, zones of upwelling
- 6.2. Low productivity in open ocean (tropics and subtropics); strong thermocline limits production
- 6.3. Each year, the sea produces 250–300 billion tons of photosynthetically produced material—consider, entire human population requires 5 billion tons of food per year

Journal Questions

1. Summary

Compare and contrast the methods by which primary productivity in the sea is measured. Then describe the factors that affect primary productivity. Note how the seasonal patterns of primary production vary according to geography. Take special note of the differences between temperate and warm seas during different seasons of the year. Consider how different regions of the ocean contribute to global marine primary production. How can predictive modeling help us to use our marine resources more wisely?

2. Define the following key words in your own words. Please feel free to add additional words to this list.

a. net primary production
b. turnover rate
c. limiting factor
d. critical depth

e. langmuir cells
f. El Niño
g. photoinhibition

3. Question Section

Write questions that show your understanding of primary production in the sea. Consider creating a question that shows the limitation of one method of measuring primary production. Do you understand the relationship between standing crop and turnover rate? Do you appreciate the role of nutrients to production rates? Can you follow the seasonal events in temperate regions and understand why primary production in the tropics shows a different pattern? Do you understand the role of satellites in measuring primary productivity and how they may help us to best use our marine resources?

4. Reflections

Consider that even though both the previous chapter and this one focus on photosynthetic organisms, learning the material in this chapter is quite different from the previous one. Reflect upon the topics of grazing and changes in photosynthetic populations in temperate regions. Identify how one can understand these concepts. What specific methods or approaches can you use to ensure that you understand the material in this chapter? Are your approaches to learn this material different from the previous chapter?

Have you visited your instructor during an office hour yet? Explain.

Practice Questions

Multiple-Choice Questions (see back of book for answers)

Match the term to its correct explanation.
1. gross primary production _____
2. standing crop _____
3. turnover rate _____

    a. the total amount of plant material produced in the sea by photosynthesis minus the portion of organic material used to sustain the lives of the photosynthetic organisms

    b. an organism divides a certain number of times per unit time

    c. an organism sinks in a spiral pattern at a certain number of revolutions per unit time

    d. the amount of plant material alive at a given time

    e. the total amount of plant material produced in the sea by photosynthesis

4. In a region of the sea where all the conditions for unlimited growth of phytoplankton exist, a graph of growth per unit time will be

    a. linear (a straight line with a constant upward slope).

    b. flat (the death of old organisms will equal the birth of new ones).

    c. in a gradual decline (because most of the new ones quickly sink into the aphotic zone).

    d. an exponential and upward curve (the number of organisms that increases per unit time always increases).

    e. all of the above, it depends upon the species.

5. In nature, what is likely to control the size of phytoplankton populations?
   a. Herbivore grazing may limit growth.
   b. A nutrient may be limiting.
   c. Light may limit growth at some times of the year.
   d. All of the above may limit growth.
   e. None of the above sets limits on phytoplankton populations.

6. How do scientists measure the height of the sea surface?
   a. Ropes are sunk into the sea with heavy weights on one end.
   b. Dark- and light-colored bottles are sunk into the sea.
   c. Radar beams from satellites can measure the height of the sea surface to 5–7 cm.
   d. All of the above
   e. None of the above

7. The dark reactions of photosynthesis occur in the
   a. grana.
   b. stroma.
   c. electron transport chain.
   d. outer smooth membrane of the chloroplast.
   e. matrix.

8. What nutrient commonly restricts the growth of diatoms?
   a. Water required for photosynthesis commonly becomes scarce during warm months in shallow seas.
   b. Carbon dioxide, essential for photosynthesis, is converted into calcium carbonate walls and skeletons and thus becomes unavailable for photosynthesis.
   c. Silica (SiO_2) often is depleted and thus limits growth (remember that diatoms have glass cell walls).
   d. Nitrogen is often not present in sufficient quantity in the sea, especially near rivers.
   e. Because we have restrictions on the use of phosphate in detergents, many marine environments near human populations do not provide diatoms with enough phosphates.

9. How does a strong thermocline affect marine primary production?
   a. Thermoclines encourage mixing of surface and deeper waters and thus provide nutrients that encourage primary production.
   b. Thermoclines prevent mixing of surface and deeper waters, which helps to keep photosynthetic organisms in the photic zone and increases photosynthetic rates.
   c. Thermoclines increase the temperature of surface waters, which increases photosynthetic rates.
   d. All of the above, it depends upon local topography.
   e. None of the above is even close!

Complete the sentences below by writing the correct answer in the blank space (see back of book for answers).

1. The type of autotroph that is responsible for most of the primary production in the sea is the

   _____ _____.

2. It is possible to have a low standing crop and high primary production if the _____ _____ is high.

3. One way to estimate standing crop is to measure the plant pigment _____.

4. When light levels are very high, photosynthesis may be inhibited because of _____.

5. The two energy-rich products of the light reaction of photosynthesis are _____ and

   _____.

6. The brown and red colors of marine plants are due to the presence of _____ pigments.

7. Coccolithophores have cells walls composed of _____ _____.
8. Process by which deep, nutrient-rich waters are brought to the surface in a specific and local area is called _____.

Chapter Outline

1.0. Introduction
    1.1.   Includes remainder of Kingdom Protista, nonphotosynthetic protozoans
    1.2.   Invertebrate phyla that dominate marine communities
    1.3.   Sequence reflects evolutionary trends
2.0. Animal Beginnings—The Protozoans
    2.1.   Kingdom Protista, not photosynthetic
    2.2.   Three phyla thrive in benthic and planktonic communities
    2.3.   Sarcomastigophora
        *   Flagella or pseudopods = = > flagellated or amoebalike
        *   Foraminiferans: shelled amoeba, planktonic, benthic, or attached
            **   Calcite or sand shells and pseudopods (movement, attachment, collect food)
            **   Globigerina ooze - chalk cliffs of Dover
        *   Radiolarians - SiO_2 skeletons
    2.4.   Ciliophora
        *   Cilia
        *   Tintinnids with vase-shaped lorica
    2.5.   Labyrinthomorpha
        *   Free-living on sea grasses and benthic algae
        *   Small cells produce a network of slime
3.0. Defining Animals
    3.1.   Boundary between kingdoms Protista and Animalia vague
        *   Colonial protozoans lack contractile muscles and signal conducting nerves
        *   Animals produce zygote that develops into hollow blastula
    3.2.   Porifera
        *   Sponges = most simple multicellular animals
        *   Flagellate choanocytes drive water (with food and O_2) through pores and channels to spongocoel and then out the osculum
        *   Attached
        *   Radial or asymmetrical
        *   Commercial sponges have spongin fibers
        *   Others have $CaCO_3$ or SiO_2 spicules
    3.3.   Placozoa
        *   Ciliated, 2–3 mm, flattened plate, one species
        *   Free-swimming in Pacific Ocean
4.0. Radial Symmetry
    4.1.   Introduction
        *   Oral-aboral axis
        *   Diffuse nerve net
    4.2.   Cnidaria
        *   Jellyfish, sea anemones, corals, and hydroids
        *   Nematocysts produced in cnidoblasts
        *   Inner and outer layers of body wall separated by mesoglea
        *   Medusae or polyps
        *   Class Hydrozoa: colonial hydroids and siphonophores (e.g., *Physalia.*)

      *   Class Scyphozoa: medusoid jellyfish

      *   Class Anthozoa: corals and sea fans; secrete $CaCO_3$.

  4.3.   Ctenophora

      *   Planktonic forms with cilia in longitudinal bands = ctenes

      *   Radial symmetry, medusalike body, tentacles with colloblast (sticky) cells capture zooplankton prey

      *   Cilia in rows called ctenes

5.0. Bilateral Symmetry

  5.1.   Platyhelminthes

      *   Many colorful bottom-living forms with cilia on underside

  5.2.   Nemertina

      *   Benthic, ribbon worms

      *   Complex organ systems with proboscis

  5.3.   Gnathostomulida

      *   Benthic, jaw worms, 1 mm, no anus

  5.4.   Gastrotricha

      *   Benthic, cylindrical, less than 1 mm

  5.5.   Kinorhyncha

      *   Similar to gastrotrichs but have segmented cuticle

  5.6.   Priapulida

      *   Benthic, wormlike, less than 10 cm, polar and subpolar

  5.7.   Nematoda

      *   Most abundant multicellular animals in benthos

  5.8.   Entoprocta

      *   Benthic, colonial, calcareous encrustations

      *   U-shaped gut with mouth and anus opening within a ciliated crown of tentacles = lophophore

6.0. The Lophophore Bearers

  6.1.   True coelom

  6.2.   Ectoprocta

      *   Benthic, colonial, encrusting

  6.3.   Phoronida

      *   Benthic, form tubes, less than 20 cm

  6.4.   Brachiopoda

      *   Lamp shells, benthic, long history

      *   Shell resembles mollusk, but not symmetrical

7.0. The Mollusks: Mollusca

  7.1.   Unsegmented, exoskeleton, and display cephalization

  7.2.   Like annelids, 4 classes have trochophore planktonic larva

  7.3.   Class Amphineura: chitins, 8 plates, radula

  7.4.   Class Gastropoda: snails, slugs, limpets, abalones, and nudibranchs

  7.5.   Class Scaphopoda: tusk shells

  7.6.   Class Bivalvia: mussels, clams

  7.7.   Class Cephalopoda: squids, octopuses, cuttlefish, and nautiluses

      *   Carnivorous predators, sucker-lined tentacles

      *   Largest invertebrates, 20 m

8.0. More Wormlike Phyla

  8.1.   Benthic, with hydrostatic skeletons

  8.2.   Sipuncula: intertidal peanut worms

  8.3.   Echiurida: benthic, with extensible proboscis

  8.4.   Pogonophora: lack digestive system; symbiotic algae provide energy needs

      *   Deep water, tube-living worms

8.5.   Hemichordata: acorn worms, found in shallow water in tubes or burrows

8.6.   Chaetognatha: arrowworms, planktonic carnivores

9.0. Segmented Animals

    9.1.   Annelida

        *   Class Polychaeta, 5,000 species, metameres

        *   Often filter feed in tubes

    9.2.   Arthropoda

        *   Segmented, exoskeleton of chitin with joints

        *   Class Merostomata: *Limulus*

        *   Class Pycnogonida: sea spiders

        *   Class Crustacea: shrimps, crabs, and lobsters

           **   Comprised of large, benthic forms and zooplankton

           **   Two pairs of antennae

           **   Subclass Copepoda: filter feeder, pelagic

           **   Euphausiids: prey of fish, whales, seals, and penguins

10.0.   Radial Symmetry Revisited: Echinodermata

    10.1.   Benthic adults with pentamerous radial symmetry, calcareous skeleton

    10.2.   Bilaterally symmetrical larval stages

    10.3.   Water-vascular system

11.0.   The Invertebrate Chordates: Chordata

    11.1.   Notochord, hollow dorsal nerve cord, and pharyngeal arches

    11.2.   Subphylum Urochordata: filter-feeding sea squirts and gelatinous salps

    11.3.   Subphylum Cephalochordata: *Amphioxus*

Journal Questions

1. Summary

Write a narrative that describes the phyla discussed in this chapter. Try to include as much detail as possible.

2. **Define the following key words in your own words.** Include the name of the phylum, which the term occurs, when appropriate. Please feel free to add additional words to this list.

   a. medusa
   b. foraminiferans
   c. colloblast cells
   d. blastula
   e. osculum
   f. radial symmetry

   g. ribbon worms
   h. lophophore
   i. coelom
   j. radula
   k. exoskeleton

3. **Question Section**

Create questions that show your understanding of the various phyla in this chapter. Try to write questions that reflect evolutionary relationships.

4. Reflections

Consider the difficulty of learning the names of the various phyla. Many of the organisms in this chapter are probably not familiar to you. How does this make learning about them difficult? Have the above writing exercises been of help to you in learning this material? Do you need to add other approaches?

Practice Questions

Multiple-Choice Questions (see back of book for answers)

1. Locomotion in many protozoans is dependent on
   a. muscle action.
   b. cell-wall contractions.
   c. cilia or flagella.
   d. all of these.
   e. none of these.

2. Two major marine protozoans are
   a. diatoms and bacteria.
   b. Cnidarians and Entoprocts.
   c. flatworms and rotifers.
   d. radiolarians and foraminifers.
   e. none of the above.

3. What do the Ciliophora and Placophora have in common?
   a. Both are unicellular.
   b. Both are covered by cilia.
   c. Both are photosynthetic.
   d. All of the above.
   e. None of the above.

4. The most simple multicellular phylum listed below is the
   a. Ciliophora.
   b. Sarcomastigophora.
   c. Porifera.
   d. Cnidaria.
   e. Annelida.

5. Water exits a sponge through a pore called the
   a. osculum.
   b. mouth.
   c. anus.
   d. spongocoel.
   e. coelom.

6. Cnidarians are armed with structures called
   a. medusae.
   b. cilia.
   c. nematocysts.
   d. lophophores.
   e. coeloms.

7. Ciliated feeding tentacles are found in the phylum
   a. Entoprocta.
   b. Lophophores.
   c. Ciliophora.
   d. Coeloms.
   e. Kinorhynchs.

Match the description to the correct phylum.
8. The rasping organ called the radula is found in what phylum? _____
9. Which phylum has segmentation and a hard exoskeleton composed of chitin? _____
10. Which wormlike animals have a notochord, a hollow dorsal nerve cord, and pharyngeal arches? _____

   a. Mollusca
   b. Arthropoda
   c. Brachiopoda
   d. Ectoprocta
   e. Chordata

11. The mantle tissue of squids and octopuses is more muscular than the mantles of other mollusks and is used primarily for
   a. food collecting.
   b. shell formation.
   c. shell collection.
   d. locomotion.
   e. reproduction.

Complete the sentences below by writing the correct answer in the blank space (see back of book for answers).

1. The tube feet of echinoderms move because they are connected to the _____ _____.
2. The type of radial symmetry of echinoderms is said to be _____.
3. An example of an animal in the class Merostomata is commonly called the _____ _____.
4. Marine polychaetes belong to the phylum _____.
5. Mollusks with an array of sense organs in their head are said to have a pattern of body organization known as _____.
6. The most abundant organisms in bottom sediments probably belong to the phylum _____.
7. Members of the Platyhelminthes are said to have _____ symmetry.
8. Ctenophores have bands of cilia known as _____.
9. Cnidarians that are sessile benthic forms are said to represent the _____ form of the life cycle.
10. Calcareous or siliceous skeletal elements in sponges are called _____.

Chapter 7 Marine Vertebrates

Chapter Outline

1.0. Introduction
    1.1.   Occupy all marine habitats; dominate pelagic zones
    1.2.   Great diversity of species, morphology, and size
2.0. Vertebrate Features
    2.1.   Basic chordate features
       *   Notochord
       *   Postanal tail
       *   Hollow nerve cord
       *   Pharyngeal pouches
    2.2.   In addition:
       *   Vertebral column
       *   Brain
       *   Myomeres
    2.3.   Amphioxus-like ancestor, 0.5 billion years ago
3.0. Marine Fishes
    3.1.   Agnatha—the jawless fishes
       *   Lack paired appendages, lower biting jaws, and skin scales
       *   Hagfishes (marine benthic scavengers) and lampreys
       *   Anadromous
    3.2.   Chondrichthyes—sharks, rays, and chimaeras
       *   Paired fins, biting jaws with teeth—effective predators
       *   High levels of urea because NaCl 50% of seawater
       *   Cartilaginous skeletons, 0.5 m to 10 m
    3.3.   Osteichthyes—the bony fishes
       *   Subclass Crossopterygii—lobe-finned fishes
          **   Coelacanth osmoregulates with urea
          **   Common ancestor with land-dwelling vertebrates
       *   Subclass Actinopterygii—ray-finned fishes
          **   Superorder Teleostei, marine, 35 orders
          **   Hypoosmotic body fluids, swim bladders, bony skeletons
4.0. Marine Tetrapods
    4.1.   Introduction
       *   Three classes of air-breathing tetrapods pelagic
       *   Hypotonic body fluids
    4.2.   Marine reptiles
       *   Snakes, turtles, iguanas
       *   Like birds, nasal glands (secrete salt) and special kidneys
       *   Excrete uric acid to reduce water loss
       *   Amniotic egg
    4.3.   Marine birds (Class Aves)
       *   Great diversity
       *   Adaptations for flight modified for the marine environment
          **   Streamlined and insulated body
       *   Vascular countercurrent heat-exchangers help prevent heat loss

4.4. Marine mammals (Class Mammalia)
   * Viviparity, hair, milk
   * No cloaca; separate openings for reproductive and digestive systems
   * Order Carnivora: seals, sea lions, walruses, sea otters, and polar bears
      ** Suborder Pinnipedia (1st 3 above), common terrestrial ancestor based upon DNA and protein studies
   * Order Sirenia: manatees, dugongs, and sea cows
      ** Paddlelike tails, herbivorous, shallow, warm waters
   * Order Cetacea (2 m – 30 m)
      ** Terrestrial ancestors
      ** Lack pelvic appendages and body hair
      ** Dorsal blowhole, streamlined, tail fluke propels
      ** Suborder Mysticeti (baleen whales), baleen
      ** Suborder Odontoceti (toothed whales)

Journal Questions

1. Summary

Write a narrative of this chapter that shows your understanding of the basic vertebrate features and detailed descriptions of the different classes of fishes and tetrapods.

2. **Define the following key words in your own words.** Please feel free to add additional words to this list.

a. myomeres   f. swim bladder
b. notochord   g. amniotic egg
c. anadromous   h. nasal gland
d. urea   i. baleen
e. coelacanth

3. Question Section

Write questions you have about the classification of marine vertebrates. Can you distinguish between classes and orders? You may wish to address evolutionary questions that you may have. For example, is it clear to you how the great whales of the sea evolved from land animals? Why did homeotherms first evolve on land and then adapt to the sea?

4. Reflections

Again we have a chapter that concerns classification and contains words and animals that may seem strange. Why is learning material of this type a special challenge? How successful have you been so far? What new insights have you developed to deal with material of this kind?

Practice Questions

Multiple-Choice Questions (see back of book for answers)

1. Vertebrates have all of the features below except for the
   a. spongocoel.
   b. notochord.
   c. postanal tail.
   d. brain.
   e. myomeres.

Match the class of fishes to their characteristics.
2. Osteichthyes _____
3. Chondrichthyes _____
4. Agnatha _____

   a. homeothermic
   b. lack lower biting jaws
   c. adult skeleton composed of cartilage
   d. have a swim bladder
   e. amniotic eggs

5. Lobe-finned fishes, including the coelacanth, belong to the subclass
   a. Osteichthyes.
   b. Crossopterygii.
   c. Actinopterygii.
   d. Actinomycin.
   e. Cetacea.

6. What taxon of marine fishes contains most of the species commonly found on our plates at dinner?
   a. Teleostei
   b. Tolstoy
   c. Squamata
   d. Squaliformes
   e. Agnatha

7. In general, the NaCl concentrations of the body fluids of marine fishes is _____, relative to seawater.
   a. great
   b. less
   c. about the same
   d. all of the above, no patterns are found
   e. none of the above

8. What class(es) of marine vertebrates is(are) homeothermic?
   a. Reptilia
   b. Agnatha
   c. Mammalia
   d. all of the above.
   e. none of the above.

9. Herbivorous marine mammals that live in warm coastal environments are called
   a. manatees.
   b. dugongs.
   c. sea cows.
   d. all of the above.
   e. none of the above.

10. A filter feeding whale is able to eat because of
    a. baleen plates.
    b. dental plates.
    c. numerous sharp teeth.
    d. comb rows.
    e. their oral disk and hook-shaped teeth.

Complete the sentences below by writing the correct answer in the blank space (see back of book for answers).

1. Every vertebrate has a _____ nerve cord located _____ to the notochord.
2. The first vertebrates were likely to have been jawless _____.
3. Fishes that feed in salt water but spawn in fresh water are said to be _____.
4. Because the concentration of NaCl in sharks is only about 50% that of seawater, the fishes concentrate _____ to achieve osmotic equilibrium.
5. The class of vertebrates called _____ has no marine members.

Chapter Outline

1.0. Introduction
    1.1.   Contains 90% animal species and most larger marine plants; live in association with sea bottom
    1.2.   Primary producers restricted to shallow near-shore areas
    1.3.   Animals from high intertidal to deep trenches
2.0. Living Conditions on the Seafloor
    2.1.   Introduction
        * Epifauna: crawl on surface
        * Infauna: live within bottom
            ** Macrofauna: swallow or displace particles as they move (e.g., clams, worms, and crabs)
            ** Microfauna: less than 50 μm.
            ** Meiofauna: interstitial animals, occupy spaces between sediment particles
    2.2.   Seafloor characteristics
        * Rain of detritus may be the only source of food
        * From solid rock to soft, loose deposits
        * Interactions of waves with coast results in greater energy directed at headland and less on coves and bays
        * Erosion of headlands results in sediments that are moved into deeper waters or into bays
        * Rocky outcrops are scattered and relatively uncommon
        * A few minerals precipitate into irregular deposits
        * Near shore, mostly small sediment particles
        * On continental shelves, mostly transported by rivers
    2.3.   Animal-sediment relationships
        * Benthic animals modify bottom, i.e., some erode rock
        * Distribution of plants and animals influenced by bottom
        * Suspension feeders, eat small plankton and detritus
            ** Require clean water to avoid clogging filters
            ** Often live on rocks
        * Deposit feeders, mudflats and deep-ocean basins
        * Absorptive feeders: tube worms and echinoderms
        * Predators and scavengers: include both fishes and birds
        * Algal grazers: snails eat kelp, sea urchins eat algae
    2.4.   Larval dispersal
        * Many benthic forms widespread; e.g., *Mytilus edulis* common on east and west temperate coasts of the United States
        * Barnacles can cross the ocean on the hulls of ships
        * Meroplankton extend geographic range of many forms; a temporary planktonic larval stage(s)
        * 75% of shallow-water benthic invertebrates produce planktonic larval stages that persist for at least 2–4 weeks
        * On appropriate bottom, larvae metamorphose into young of bottom stage
        * Some spawn all year, other have seasons, often in response to temperature

* Reproductive success demands that fecundity exceeds mortality, e.g., *Aplysia,* 478 million eggs in 5 months of laboratory observations
* Broadcast spawners, many common forms; pheromones cause sperm to be attracted to eggs (e.g., Echinoderms)

3.0. Intertidal Communities

3.1. Introduction

* Total biomass m² at low tide = 10X bottom at 200 m and 3,000X abyssal areas
* Littoral, or intertidal, zone caused by rise and fall of tides
* Low tide = physiological stress
* Osmotic problems and predators (i.e., birds and raccoons)
* Breaking waves impose large forces
* Physical + biological factors + biological role = niche

3.2. Rocky shores

* A grazing food chain
* The upper intertidal
  ** Conditions from terrestrial to marine
  ** Cyanobacteria: *Calothrix* or lichen *Verrucaria*
  ** Tufts of *Ulothrix*
  ** Few snails, limpets, or crustaceans graze here
  ** *Littorina* is an air-breather
  ** Below blue-greens, barnacles; feed a few hours per month at spring high tides
* The middle intertidal
  ** Green, red, and brown algae - food for grazers
  ** Tide pools protect hermit crabs, snails, anemones, and fish
  ** Mussels, barnacles, chitons, *Fucus*
  ** Limited by solid substrate and nutrients in water
  ** Most animals have free-swimming larval stages
  ** Barnacles tend to outcompete *Fucus*
  ** Barnacles are eaten by sea stars, snails, and fish
  ** Mussels attach to rocks, algae, or barnacles by byssal threads; eventually outcompete barnacles
  ** Mussels eaten by sea stars, snails; gradually new space is cleared by wave action and predators
  ** Submussel habitat: clams, worms, shrimps, crabs, hydroids, algae = climax community created by biological succession
* The lower intertidal
  ** Brown, red, and green seaweeds form a canopy
  ** East Coast, *Metridium.* Some anemones preyed on by snails and sea spiders
  ** Echinoderms: sea stars, sea urchins, brittle stars, and sea cucumbers; sensitive to desiccation and changes in salinity. Sea stars eat mussels, barnacles, snails, anemones, and other echinoderms

3.3. Sandy beaches and muddy shores

* Detritus food chains dominate
* Beaches made of quartz grains, volcanic sand, or pulverized carbonate skeletons. Require waves gentle enough not to wash sediment away but strong enough to wash away fine silts and clays
* Mudflats, found in estuaries or quiet bays
  ** Interstitial space size determines O_2
  ** Bacteria and fungi live in anaerobic muds
  ** Gastropods, bivalves, crustaceans, and polychaete worms

* Sandy beaches seem desolate but support a rich fauna
  ** Shifting sands mean all permanent residents lie below the surface
  ** Amphipods, crabs, isopods, and worms common
  ** Fiddler crabs emerge at low tide - biological clock based upon the tides - circalunadian rhythm (24.8 hours)
  ** Fiddler crab coloration—darken during day and lighter at night - circadian rhythm (24 hours)
* During high tide, submerged sandy beaches and mudflats visited by crabs, shrimp, and fish. At low tide, birds, bats, rats, raccoons, and coyotes patrol the shores
* At low tide, many sand-living organisms switch from aerobic to anaerobic respiration

Journal Questions

1. Summary

Write an extensive narrative describing how each type of seafloor has a flora and fauna that characterize its specific niche. Carefully note how vertical zonation results in regions with specific life-forms. Try to imagine the difficulties of living in each environment.

2. **Define the following key words in your own words.** Please feel free to add additional words to this list.

a. microfauna
b. interstitial animals
c. suspension feeder
d. deposit feeder
e. meroplankton

f. fecundity
g. broadcast spawner
h. pheromone
i. niche
j. byssal threads

3. **Question Section**

This is a long and complex chapter that is likely to raise many questions. Do you understand the relationships between the conditions on the bottom and the forms of life found? Do the reproductive strategies of benthic organisms make sense to you? Do you understand the details of vertical zonation on rocky shores? Can you understand why oxygen levels on sandy beach communities and mudflats differ?

4. Reflections

In this chapter we hope to learn some of the relationships between organisms and their environment. Is it difficult to appreciate the different challenges presented by rocky shores, sandy beaches, and mudflats? Pretend you are one of the organisms described in the chapter, and write a brief diary of a typical day in your life.

Practice Questions

Multiple-Choice Questions (see back of book for answers)

Match the type of benthic organism to its appropriate description.
1. interstitial animals _____
2. macrofauna _____
3. epifauna _____

    a.   benthic animals larger than 0.5 mm that either swallow or displace the sediment particles around themselves as they move

    b.   small organisms that move with the water currents

    c.   large animals that can swim against strong currents

    d.   crawl over or attach to surface of sea bottom

    e.   intermediate-sized organisms that live in the spaces between sediment particles

4. Epifauna and large benthic plants are most likely to live on
    a.   a firm or solid bottom.
    b.   a sandy bottom of a sloping beach.
    c.   a mudflat.
    d.   all of the above are equally likely.
    e.   none of the above.

5. Animals that live in or on the surface of deep-ocean basins are most likely to be
    a.   suspension feeders.
    b.   deposit feeders.
    c.   withdrawal feeders.
    d.   predators of small fish.
    e.   all of the above.

6. Slow-moving benthic organisms can be widely distributed if they
    a.   produce meroplankton.
    b.   crawl in a straight line.
    c.   live very long lives.
    d.   all of the above.
    e.   none of the above.

7. What is(are) the possible advantage(s) to a benthic species to produce planktonic larvae?
   a. The young do not compete directly with adults for scarce resources.
   b. Adults increase the likelihood of large numbers of offspring surviving.
   c. If the environment of the adults becomes unfavorable, their offspring may survive in a distant location.
   d. All of the above
   e. None of the above

8. Reproductive success demands that
   a. mortality exceeds fecundity.
   b. fecundity exceeds mortality.
   c. ontogeny recapitulates phylogeny.
   d. all of the above.
   e. none of the above.

9. Eggs and sperm of broadcast spawners are likely to fertilize because
   a. of the large numbers of eggs and sperm, haphazard meetings are common.
   b. pheromones cause some organisms to release their gametes at the same time.
   c. ontogeny does not recapitulate phylogeny.
   d. all of the above.
   e. none of the above.

10. Tidal fluctuations of sea level cause intertidal organisms to experience
   a. dry and hot weather.
   b. wave shock.
   c. rainfall and freshwater runoff.
   d. all of the above.
   e. none of the above.

Complete the sentences below by writing the correct answer in the blank space (see back of book for answers).

1. The southern tip of Florida is our only tropical shoreline; it is characterized in places by shrubby plants known as _____.
2. Energy transfer on a typical rocky shore involves a _____ food chain while sandy shores rely more on _____.
3. The dark regions on the upper intertidal zone are caused by lichens or members of the _____.
4. A common genus of snail in the upper intertidal zone that can breathe air is _____.
5. In the middle intertidal zone, barnacles and algae may be replaced by _____.
6. The changes described in the previous question are an example of _____ _____.
7. Anaerobic microbes in fine-grained muds may reduce sulphate to _____ and thus cause a memorable odor.
8. Fiddler crabs are darker during the day compared to the night because of a _____ _____.

Chapter Outline

1.0. Introduction
    1.1.   Semienclosed coastal embayments - freshwater rivers meet the sea
    1.2.   Molded by suspended particles and tide and current patterns
    1.3.   Unstable because of physical, geological, chemical, and biological factors
    1.4.   Change from fresh to salt water may extend from a few hundred meters to many miles
    1.5.   From shallow and flat bays to steep-sided fjords
    1.6.   Salt water moves upstream during high tide—causes sediment load with pollutants to be deposited near mouth of river
    1.7.   Very productive - turbulent mixing - 2/3 of fish catches depend on estuaries as feeding area and/or nursery
    1.8.   Becoming an endangered natural habitat; economic exploitation (i.e., development and place to dispose of wastewaters) and pollution (i.e., pesticides)
2.0. Types of Estuaries
    2.1.   Most formed by glacial erosion - sea level was 150 m lower
    2.2.   Coastal plain estuaries
        *   Broad, shallow embayments
    2.3.   Drowned river valley estuaries
        *   Caused by rising sea levels; constantly modified by erosion
    2.4.   Bar-built estuaries
        *   Near-shore sand and mud moved by coastal wave action to build an obstruction or bar, e.g., Gulf of Mexico
        *   Restricted mouths to fan-shaped deltas
        *   Fjords deeper upstream; sills at mouth may lead to stagnant conditions in the bottom of deep fjords
3.0. Estuarine Circulation
    3.1.   Salinity may increase from the surface down. Why?
    3.2.   Higher density and nutrient-rich seawater may move in near the bottom of an estuary and cause an upwelling that promotes high productivity
    3.3.   Flushing time may vary from days to years
4.0. Salinity Adaptations
    4.1.   Benthic forms must be able to tolerate changes in salinity and internal osmotic pressure
    4.2.   Most forms derived from marine organisms
    4.3.   Some adapt by closing their shells - anaerobic respiration
    4.4.   Others bury into mud
    4.5.   Tunicates, anemones - osmotic conformers
    4.6.   Many crustaceans are osmoregulators
    4.7.   Most stenohaline - limited range of salinity
    4.8.   Few euryhaline
5.0. Sediment Transport: Creating Habitats
    5.1.   Erosion and the life history of the river determine sediments
6.0. Estuarine Habitats and Communities
    6.1.   Wetlands - highest elevation - covered at high tide - dense plant communities
    6.2.   Mudflats exposed at low tides
    6.3.   Channels, always under water

6.4. Temperate wetlands: salt marshes
* Dominated by halophytes, plants tolerant to salt water
* Develop in muddy deposits around edges of temperate and subpolar estuaries
* May have several plant species:
** Salt grass and pickelweed may grow in lowest portion; stores excess salt in fleshy leaves
** Higher zones - plants that cannot be submerged for long periods of time
* Important base of estuarine food web - most enters as detritus - then to bacteria

6.5. Tropical wetlands: mangals
* Salt-tolerant shrubby trees with prop roots in black anaerobic mud
* Taller (up to 25 m) mangroves of south Florida especially prone to hurricane damage

6.6. Mudflats
* Where marine waters mix with rivers - flocculant adds to bottom mud
* Anaerobic with abundant bacteria - hydrogen sulfide
* Clams, mud shrimp, worms, and other burrowing animals
* Primary producers = diatoms, larger algae, and eel grass.
* Eel grass (a nutrient pump) feeds ducks, invertebrates, fish, and insect larvae

6.7. Channels
* Plankton, fish larvae, and invertebrates inhabit channels
* Herring and sole spawn here
* Crabs use it as a nursery
* Salmon may linger a while and feed before they go upstream to spawn

7.0. Economic Uses of Estuaries
7.1. Resting and feeding stop for migratory birds such as ducks and eagles
7.2. A natural filter that can trap and render pollutants harmless
7.3. Moderate flooding
7.4. Most have been modified; most seaports are made in estuaries
7.5. 75% in United States have been lost due to dredging, draining, and diking
7.6. Pesticides often in high concentration, especially PCBs and DDT

8.0. The Chesapeake Bay System
8.1. Five major and smaller estuaries, linked when Susquehanna River valley flooded after LGM
8.2. Human activities - heavy metals, pesticides, sewage
8.3. Still, a protein factory - blue crabs, oysters, striped bass, etc.
8.4. Up-Bay, low salinity; mid-Bay, brackish; lower-Bay, marine
8.5. Health of Bay threatened; 250X increase in cyanobacteria and dinophyte concentrations since 1950

Journal Questions

1. Summary

Our narrative will focus on one critical environment—the estuary. Include a description of the various types of estuaries. What is the origin of the estuary? Consider the circulation of water that results in unique patterns of salinity. What strategies do different organisms engage to meet the salinity challenges? Review the details of specific habitats: the salt marsh, the mangal, the mudflat, and the channel. Why can the Chesapeake Bay system serve as a model estuary? How have human activities affected this environment?

2. **Define the following key words in your own words.** Please feel free to add additional words to this list.

   a. coastal plain estuary   e. stenohaline
   b. fjord   f. wetlands
   c. isohalines   g. mangal
   d. osmotic conformer   h. channel

3. **Question Section**

Do you understand how and why salinity changes in the estuary? Is it clear why most estuaries in the northern regions of the United States are of relatively recent origin? How does a mudflat differ from a channel? Why do the states of Maine and Washington lack mangal communities?

4. **Reflections**

Why is an estuary a complex environment? Does that make it difficult to study? Can you see how the role of the mangal in south Florida is similar to the rocky shore of New England?

Practice Questions

Multiple-Choice Questions (see back of book for answers)

1. Estuaries are places where
   a. some of the most productive marine systems are found.
   b. fresh water mixes with salt water.
   c. human impact is greatly felt.
   d. all of the above.
   e. none of the above.

Match the type of estuary with its description.
2. Coastal plain estuary _____
3. Bar-built estuary _____

   a. a fan-shaped structure forms at the mouths of rivers that carry a great deal of sediment
   b. formed when melting ice of the last glacial period flooded V-shaped channels
   c. very deep estuary with a shallow sill at the mouth
   d. where sand and mud are moved to form a long obstruction
   e. all of the above

4. Most animals that live in an estuary are
   a. stenohaline.
   b. euryhaline.
   c. isohaline.
   d. all of the above; they are found in equal numbers.
   e. none of the above.

5. A habitat that is always under water is a
   a. channel.
   b. wetland.
   c. mudflat.
   d. all of the above.
   e. none of the above.

Complete the sentences below by writing the correct answer in the blank space (see back of book for answers).

1. The time for water in an estuary to move out to sea is called the _____ _____.
2. Animals, like tunicates, that cannot control their osmotic state are said to be _____ _____.
3. In the mangal, the dominant plant is the _____.

Chapter Outline

1.0. Tropical CaCO$_3$ reef structure is the basis for diverse marine communities
2.0. Reef-Forming Corals
    2.1.   Cnidarians, radially symmetrical, nematocysts, attached
    2.2.   Coral polyps sit in calcareous skeletal cups = corallites
          *   Septa from the center of each corallite
          *   Polyp can secrete a new elevated partition
          *   Asexually bud off new polyps
          *   Sexual reproduction - planktonic larvae
    2.3.   Corals ubiquitous, reef-formers - tropic and subtropic (18$^{+\circ}$ C)
    2.4.   Indian and Pacific oceans, 700 species
    2.5.   Atlantic Ocean, 35 species
    2.6.   Need clean water, firm sea bottom, high salinity, sunlight. Thus, do not thrive near mouths of rivers
    2.7.   Symbiotic zooxanthelle - unicellular dinophytes - 1 million per cm^2 of surface area
          *   90% of organics produced by photosynthesis transferred to coral tissue = 3X phytoplankton production over reef
    2.8.   Large polyps also eat small fish and large zooplankton
    2.9.   Small polyps eat small plankton and detritus.
    2.10.   Most corals can also harvest bacteria, fish slime, and organic substances in the water
    2.11.   Warm water needed for high rates of CaCO$_3$ deposition
    2.12.   Fringing reefs
          *   Borders along shorelines, e.g., Hawaiian reefs
    2.13.   Barrier reefs
          *   Further offshore and separated by a lagoon
          *   Great Barrier Reef - 2,000 km
    2.14.   Atolls
          *   Ring-shaped with a few islands
    2.15.   Development of atolls from volcanic features was first described by Darwin
    2.16.   Darwin was unaware of fluctuations in sea level
          *   17,000 years ago during the last glacial maximum (LGM), sea level was 150 m below the present level. Ice melted and sea level rose 1 cm per year. Those coral reefs that did not rise fast enough died—they went past the "Darwin point"
    2.17.   Coral reefs stand in sharp contrast to the unproductive tropical seas; trophic relationships largely unknown
3.0. Reproduction in Corals
    3.1.   Asexual: bud new polyps or broken into clonal fragments by wave action
    3.2.   Sexual: 65% broadcast spawners—many on the same night
    3.3.   Zygote ---> planula larvae ---> settle and develop into adults if conditions are correct
    3.4.   Larvae may travel quite a distance to form new coral reefs
4.0. Zonation on Coral Reefs
    4.1.   Depends upon wave force, water depth, temperature, salinity, etc.
    4.2.   150–50 m deep on reef slopes - a few fragile species
    4.3.   50–20 m deep - below wave action - transition zone
          *   Many delicate forms

4.4. 20 m - below low-tide mark - zone of spurs or buttresses
* Dominated by massive coral growths and encrusting corraline algae; many fish
4.5. Reefs are affected by trade winds and the waves that they create
* Algal ridge of calcareous algae results
4.6. Then a reef flat, shallow; microatolls where water 1 m
4.7. Many animals including giant clam, *Tridacna*
* Over a meter across with zooxanthellae in mantle tissue
4.8. Lagoon reef
* Shallow margin
* Leeward - free of wave action
* Luxuriant coral, crustaceans, echinoderms, mollusks, etc.

5.0. Coral-Reef Fishes
5.1. Fish in reef and lagoon, thousands of species
* Moray eels, porcupine fish, butterfly fish, groupers, gobies, sea horses, sharks, etc.
5.2. Symbiotic relationships
* Pilot fish and remoras commensal with sharks
* Mutualistic relation between anemonefish and anemones
* Cleaning symbiosis
* Many parasites: viruses, bacteria, flatworms, roundworms, and leeches

6.0. Coloration in Coral Reef Fish
6.1. Internal pigments and iridescent surface features
6.2. Most have color vision
6.3. Concealment, disguise, and advertisement
6.4. Some change color by expanding and contracting chromatophores
6.5. Other cells contain iridocytes—crystals that produce a range of colors
6.6. A disguised eye is a protected eye
6.7. Colors play a role in breeding
6.8. Fish in only warm waters are brightly colored. Why?

Journal Questions

1. Summary

Review in detail the biology of the corals that form coral reefs. Note carefully how primary productivity is different here compared with other environments. Include a description of the different types of reefs and consider the origin of each type. What is the fate of a reef when sea level rises or falls? Note the zonation on coral reefs. You may wish to compare this with zonation on a rocky shore. Describe the unique symbiotic relationships and coloration seen in reef animals.

2. Define the following key words in your own words. Please feel free to add additional words to this list.

a. corallites
b. zooxanthellae
c. fringing reef
d. Darwin Point
e. planula larvae

f. algal ridge
g. remora
h. cleaning symbiosis
i. chromatophore

3. Question Section

Make sure you understand the structure and reproduction of corals. Are you clear about the origin of coral reefs and how changes in sea level may affect them? Do you understand why reefs do not occur near the mouths of tropical rivers? Why do reef fishes have bright colors and stripes while fishes in temperate waters appear quite dull?

4. Reflections

Like tropical forests of the land, the coral reefs contain enormous species diversity. What is it about the reef that is so unique? Consider the primary producers in your response. Do the unusual features of the biology of coral make this subject difficult to learn?

Practice Questions

Multiple-Choice Questions (see back of book for answers)

1. The basic structural units of a coral are
   a. polyps.
   b. corallites.
   c. septa.
   d. medusae.
   e. nematocysts.

2. To grow well, corals need
   a. clean water.
   b. firm sea bottom.
   c. plenty of sunlight.
   d. all of the above.
   e. none of the above.

3. What do the corals gain from the symbiotic zooxanthellae?
   a. a constant, protected environment
   b. nutrients like CO^2
   c. dull gray colors that help the coral avoid predators
   d. all of the above
   e. none of the above

4. A ring-shaped reef with a few islands projecting above sea level is called a(n)
   a. atoll.
   b. fringing reef.
   c. barrier reef.
   d. septa.
   e. zooxanthellae.

5. After the last glacial maximum
   a. sea levels fell, and coral reefs increased in size.
   b. sea levels fell, and most coral reefs died.
   c. sea levels rose, and coral reefs grew larger as a result.
   d. sea levels rose, and only those corals that could grow at the same rate as changes in sea levels survived.
   e. sea levels remained the same, and thus the distribution of coral reefs has not changed in 17,000 years.

6. At what depth in the sea do most of the delicate coral species live?
   a. 150 m below sea level
   b. 150–50 m below sea level
   c. 50–20 m below sea level
   d. 20–0 m below sea level
   e. the intertidal zone

7. The relationship between the anemonefish and several species of sea anemones is likely to be
   a. parasitic.
   b. predaceous.
   c. commensal.
   d. mutualistic.
   e. anomalous.

8. In fish, cells that contain reflecting crystals of guanine and produce a spectrum of colors are called
   a. iridocytes.
   b. chromatophores.
   c. chromatids.
   d. all of the above.
   e. none of the above.

Complete the sentences below by writing the correct answer in the blank space.

1. The type of reef that forms borders along shorelines like the Hawaiian Islands is called a _____ reef.
2. Barrier reefs are separated from the shore by a _____.
3. More than half of reef-forming coral sexually reproduce as _____ _____.
4. Sexual reproduction in corals results in a planktonic and swimming _____ larva.
5. Corals do not live below 150 m because there is not enough _____.

58

6. In addition to coral, another large animal (more than 1 m) that also lives on the reef in a symbiotic relationship with zooxanthellae is the _____.

7. The relationship between animals whereby one eats parasites off the other is said to be a _____ symbiosis.

Chapter Outline

1.0. Introduction
     1.1. Sand and mud dominate the seafloor
     1.2. On continental shelves, level, soft bottom where life depends upon rain of plankton and detritus
     1.3. In deep ocean basins, 1 cm of ooze forms every 1,000 years
2.0. Shallow Subtidal Communities
     2.1. Recurring communities of infauna, few macrofauna
          * Worldwide similarities in cold and temperate regions where soft muddy sediments
     2.2. Rocky areas dominated by brown algae—kelp "forests"
          * *Macrocystis, Nereocystis,* and *Laminaria*
          * More niches than soft bottoms
          * Smaller brown and red algae live in the shade of the larger browns
          * High rates of primary productivity
3.0. The Abyss
     3.1. Three-fourths of the ocean floor lies below 3,000 m
     3.2. A rigorous, constant, sunless environment
     3.3. Bottom = soft, fine-grained clays with skeletons of plankton
     3.4. 2° C - below 0° C
     3.5. Pressures 300 - 600 atmospheres
     3.6. Lowered metabolic rates
     3.7. O_2 decreases with depth until O_2 minimum zone and then increases
     3.8. Energy transfer from the sea surface to the sea bottom
          * All food from above. Why?
          * Sinking rates depend upon size
          * Phytoplankton may aggregate or become compact fecal pellets that sink quickly
          * Small particles, lose most of their nutritive value
     3.9. Inhabitants of the deep-sea floor
          * Dominant forms: echinoderms (sea cucumbers and crinoids), polychaete worms, pycnogonids, and isopod and amphipod crustaceans dominate
          * Few bivalves and sea stars
          * Variety of species at least equal to inshore communities
           ** Enormous areas with few barriers to dispersal
          * Fed by fallout of feces and occasional sinking dead animals
          * Most are infaunal deposit feeders; 30–40% organic material absorbed by bacteria. Even the shed exoskeletons of crustaceans are used by the bacteria
          * Cropper: merged role of predator and deposit feeder. Reduces competition for food and permits coexistence between species sharing the same food
          * Very few produce planktonic larvae. Why?
          * Eggs are large to provide larvae with food
4.0. Deep-Sea Hot Springs
     4.1. Tectonic spreading on ridge and rise system = vulcanism
     4.2. Molten magna - basalt - forms new floor
     4.3. 1977, deep-sea hot springs first observed
     4.4. Large mussels, clams, giant tube worms, and crabs
     4.5. Seawater circulates in and out of cracks and temperature rises from 2° C to 350° C

4.6. Complex chemistry: $SO_4^{-2} ==>$ H_2S combines with metals $==>$ iron sulfide + H_2S == "black smokers"

4.7. Bacteria: $H_2S + O_2 --->$ SO_4 + energy = chemosynthesis

4.8. Bacteria eaten by clams and mussels

4.9. The giant red-plumed tube worm, *Riftia,* uses internal bacterial symbionts to oxidize the H_2S in its unique trophosome

4.10. Other chemosynthetic communities exist
    * Methane seeps and earthquake-disturbed sediments

Journal Questions

1. Summary

Write an extensive summary of subtidal communities dominated by brown algae and the communities in the deep sea. Note similarities and differences between these different strategies for life in the sea. Note the role of primary productivity in each type of environment.

2. **Define the following key words in your own words.** Please feel free to add additional words to this list.

   a. oceanic ooze
   b. epifauna
   c. cropper

   d. chemosynthesis
   e. *Riftia*
   f. whalefall

3. **Question Section**

Life below the tides may evoke many questions. The deep-sea communities are especially strange, and some of this material may be difficult to understand. For example, do the chemical reactions of the deep-sea hot springs make sense to you?

4. Reflections

In this chapter we describe environments with which you may have been totally unfamiliar with before. We explore the kelp forest, the abyss, and the deep-sea hot springs. Does the exotic nature of this material make it more difficult or more easy to learn? Does the chemistry of the hot springs present challenges? How can you best comprehend each of these different communities and the connections between them?

Practice Questions

Multiple-Choice Questions (see back of book for answers)

1. Life on the bottom of the sea largely depends upon
   a. nutrient runoff from major rivers.
   b. direct use of solar energy.
   c. the slow rain of plankton, feces, and detritus.
   d. all of the above; they are equally important.
   e. none of the above.

2. The abyss (below 3,000 m) includes _____ of the ocean bottom.
   a. 10%
   b. 25%
   c. 50%
   d. 75%
   e. 90%

3. Below 3,000 m, water averages ____° C in temperature.
   a. -50
   b. 2
   c. 25
   d. 50
   e. 75

4. What type of organisms dominate the deep-sea floor?
   a. sea cucumbers, crinoids, polychaete worms
   b. bivalves
   c. sea stars
   d. all of the above
   e. none of the above

5. The dense color of "black smokers" is due to
   a. SO_4.
   b. H_2S.
   c. iron sulfides.
   d. basalt.
   e. hemocyanin.

6. Bacteria in the deep-sea hot springs obtain energy by chemosynthesis from what reaction?
   a. $H_2O + CO_2 \rightarrow C_6H_{12}O_6 + O_2$
   b. $C_6H_{12}O_6 + O_2 \rightarrow CO_2 + H_2O$
   c. Sulfate $(SO_4^{-2}) \rightarrow H_2S$
   d. $H_2S \rightarrow$ Sulfate (SO_4^{-2})
   e. all of the above

7. Animals that feed off the bacteria in the deep-sea hot spring communities are
   a. clams.
   b. mussels.
   c. red-plumed giant tube worms.
   d. all of the above.
   e. none of the above.

Complete the sentences below by writing the correct answer in the blank space (see back of book for answers).

1. The communities on rocky outcrops in shallow water are dominated by _____.
2. With respect to species diversity, deep-sea communities have _____ species compared to shallow soft-bottom areas.
3. An animal that merges the role of predator and deposit feeder is known as a _____.
4. When sulfate ions are heated by hot basalt, they are converted into _____.
5. In deep-sea hot spring communities, *Riftia* is able to compete with bacteria for oxygen because it contains the pigment _____.

Chapter 12   The Zooplankton

Chapter Outline

1.0. Introduction
    1.1.   3-D, nutritionally dilute world
    1.2.   Concentrate in the photic zone
2.0. Zooplankton Groups
    2.1.   Meroplankton: temporary larval stages of invertebrates and fish; concentrate over continental shelves, reefs, and estuaries
    2.2.   Holoplankton: 5,000 species in two kingdoms
        *   Unicellular; cnidarians, ctenophores, mollusks, chaetognaths, crustaceans, and invertebrate chordates (tunicates)
        *   Often have high surface/volume ratios
        *   Siphonophores have pneumatophores
        *   Float + sail = worldwide distributions of *Physalia*
        *   Intestinal gases used by snail - lives near surface
        *   Neuston: live near air-water interface
    2.3.   Gelatinous zooplankton
        *   Jellyfish, siphonophores, pelagic mollusks, ctenophores, and tunicates
        *   95% water, size ranges from 1 mm–3 m
    2.4.   Crustaceans: most numerous and widespread holoplankton
        *   Copepods, euphasiids, amphipods, decapods, and ostracods
        *   Life cycles involve several stages
           **   Planktonic mysids: nauplius, protozoea, zoea, and adult
           **   Planktonic copepods: nauplius, after 6 molts --> copepodite, after 5 more molts --> adult
3.0. The Pelagic Environment
    3.1.   A realm with few obvious niches
    3.2.   Epipelagic zone: upper 200 m = photic zone
    3.3.   Climatic zones influence distribution
    3.4.   Local distributions patchy
4.0. Vertical Migration
    4.1.   Cannot make directed long-range movements
    4.2.   Often can move a few tens of meters vertically
    4.3.   Mesopelagic zone: below 200 m to 1,000 m
    4.4.   Depends upon rain of debris and fecal pellets
    4.5.   Low light makes it difficult for predators
    4.6.   Cold water lowers metabolic rates and need for oxygen
    4.7.   Cold water is more dense and thus inhibits sinking
    4.8.   Many animals live in this deep zone during the day and rise into the epipelagic zone at night
    4.9.   Euphausiids, small fish, and siphonophores often have a daily vertical migration; observed as deep scattering layers
    4.10.   Vertical migrations can also be seasonal. The copepod *Calanus* spends the winter at 1,000 m; in the spring, the larvae molt to the adult form and move into the photic zone to feed on the bloom of diatoms
5.0. Feeding
    5.1.   Chaetognaths are predators of other zooplankton, including young fish and copepods

5.2. Copepods filter feed upon particles from bacteria to large centric diatoms

5.3. Gelatinous herbivores rely on nets or webs of mucus

5.4. Number of steps in food chain depends upon availability of primary producers

Journal Questions

1. Summary

Write an extensive narrative to describe the variety of zooplankton. Consider in detail the unusual features of their environment and the life-styles that have evolved to exploit their marine world. Carefully note the events of vertical migration and feeding.

2. **Define the following key words in your own words.** Please feel free to add additional words to this list.

    a. meroplankton vs. holoplankton
    b. pneumatophore
    c. neuston
    d. nauplius

    e. epipelegic vs. mesopelagic zones
    f. biantitropical distribution
    g. deep scattering layers
    h. *Calanus*

3. **Question Section**

Develop questions concerned with zooplankton. Is the diversity in this group clear to you? Do you understand the events of vertical migration? Are the biological advantages of vertical migrations and living in dim waters clear to you?

4. Reflections

In this chapter we explore the biology of very small animals. What is it about their world and life-styles that is especially foreign to you? Do the varieties in form and life-style make this a difficult chapter to learn? Can you see the relationship between the world of the very small and the health of larger animals, such as fishes, whales, and people?

Practice Questions

Multiple-Choice Questions (see back of book for answers)

1. An animal that has a part of its life cycle as a zooplankton form is said to be a
   a. meroplankton.
   b. holoplankton.
   c. phytoplankton.
   d. all of the above.
   e. none of the above.

2. How do zooplankton remain in the photic zone?
   a. Their small size increases frictional resistance.
   b. Some secrete gases into a structure.
   c. Some have body densities similar to seawater.
   d. all of the above
   e. none of the above

3. A typical sequence of the life cycle for a typical planktonic mysid is
   a. zoea --> nauplius --> protozoea --> adult
   b. nauplius --> protozoea --> zoea --> adult
   c. nauplius --> zoea --> protozoea --> adult
   d. all of the above can happen; life cycles in this group can be very diverse
   e. none of the above

4. The biological advantage(s) to zooplankton spending part of their day in very dim light is(are)
   a. zooplankton cannot sleep in bright light.
   b. predators are especially vicious in dim light, and thus zooplankton that can avoid predation are naturally selected.
   c. decreased depth increases pressure and the metabolic rate of the zooplankton.
   d. all of the above.
   e. none of the above.

5. Describe the food chain in the Antarctic upwelling system.
   a. low in biomass but maintain tuna-sized predators
   b. relatively large primary producers, few trophic levels, supports the largest animals on earth
   c. relatively small primary producers, many trophic levels
   d. all of the above are found—it depends on the exact position
   e. none of the above

Complete the sentences below by writing the correct answer in the blank space (see back of book for answers).

1. Zooplankton that live near the air-surface interface are called _____.
2. Zooplankton that spend part of their day in the mesopelagic zone and move upward at night are said to have a pattern of _____ _____.
3. One can sometimes detect movements of animals with sound pulses and record the movements as _____ _____ _____.
4. Gelatinous herbivores capture food particles with nets or webs made of _____.

Chapter Outline

1.0. Introduction
    1.1.   5,000 nektonic animals
    1.2.   Body size crucial - 3^+ cm
    1.3.   Invertebrates: squid and few species of shrimp
    1.4.   Most fish
2.0. Vertical Distribution of Nekton
    2.1.   Epipelegic nekton: mostly carnivorous predators
        *   Effective swimmers
        *   Larger than zooplankton
        *   Able to detect and orient to prey and navigate or migrate
        *   Most lack bright colors
        *   Countershading: dark dorsal surface, silvery undersides
    2.2.   Mesopelagic zone
        *   Usually smaller than 10 cm
        *   Well-developed teeth and mouths
        *   Large, light-sensitive eyes
        *   Black at depths
        *   Photophores, light producing organs - species identification?
    2.3.   Below 1,000 m, all light from photophores
        *   Lights lure prey, illuminate blackness, recognition
        *   Oversize mouths
3.0. Getting Oxygen
    3.1.   Introduction
        *   All vertebrates active, need O_2, use hemoglobin
        *   Air-breathers at surface find air 21% oxygen
        *   Fish must use gills, seawater just a few ppm oxygen
    3.2.   Breath-hold Diving in Marine Mammals
        *   From 30–60 seconds to 30 minutes or more
        *   Marine mammals inhale and exhale very rapidly (less than 1 second for smaller dolphins)
            **   Extensive elastic tissue in lungs and diaphragms
            **   May extract 90% oxygen compared to humans at 20%
        *   Lungs collapse during deep dive, prevents N_2 uptake
        *   Extra blood volume and red blood cells with extra hemoglobin
        *   Muscles and other organs tolerate anaerobic conditions
        *   Heartbeat rate slows down (bradycardia) - 20–50% predive rate in cetaceans - diving reflex
    3.3.   Fish gills
        *   Each gill arch supports a double row of gill filaments. Each filament bears secondary lamellae associated with microscopic capillaries
        *   Gill surface may = 10X body surface
        *   Countercurrent system enhances O_2 absorption up to 85% of O_2 extracted and combined with hemoglobin

4.0. Buoyancy
- 4.1. Introduction
  - * Specific gravity of muscle = 1.05; bone, scale, or shell = 2.0; cartilage = 1.1; fat, wax, or oil = 0.8–0.9
  - * Many store fats or oils; e.g., whale = 18% blubber. Many sharks 1/4 weight = oil
- 4.2. Gas inclusions
  - * Air = 0.1% as dense as water at sea level ==> a little air can provide a lot of lift. But air is compressible—1 atm for every 10 m in depth
  - * Rigid buoyancy tanks
    - ** Cephalopods like *Nautilus* have transverse septa, partitions chambers of the shell. Chambers connected by siphuncle
    - ** Implosion depth = 500 m, do not live below 240 m
  - * Swim bladders
    - ** Most bony fish have body densities 5% greater than seawater
    - ** Internal swim bladders with nitrogen and oxygen
    - ** Develop from outpouching of esophagus
    - ** Pneumatic duct connects esophagus and swim bladder in early stages of all bony fish—physostomous condition
    - ** In most adult marine fish, physoclist swim bladder develops as the duct disappears
    - ** Bladders lacking in bottom fish and in active, moving fish such as tuna
    - ** Gas for bladder comes from gills
    - ** Volume decreases with depth and pressure, so some physoclists use gas gland to maintain volume as needed
      - *** Deep-water fish have rete mirabile—capillaries that help deliver more oxygen to gas gland
    - ** Volume must decrease during ascent, physoclist fishes use oval body to reabsorb gases
    - ** For some that live very deep, fat-filled swim bladder

5.0. Locomotion
- 5.1. Introduction
  - * Swimming has a relatively low cost-of-transport in joules of energy per kg of body weight per km traveled
  - * Sprinting (barracuda), fine maneuvering (butterfly fish), high-speed cruising (tuna) = reflect ecological niche
- 5.2. Body shape
  - * Varied if camouflage more important than speed
  - * Shape of fish influenced by
    - ** Frictional drag = sphere ideal shape
    - ** Form drag = cross-sectional area head on
    - ** Turbulence = length 4.5X greatest diameter + blunt at front end + point in rear
  - * Rapidly accelerating fish are thinner and may be recognized by prey
  - * Maneuvering fish are more oval and use additional drag for fine positioning adjustments
- 5.3. Fins
  - * Thrust comes from bending of myomeres throwing body into curve or wave
- 5.4. Caudal fins
  - * Flare dorsally and ventrally
  - * Increase surface area available to provide thrust but also increase frictional drag
  - * Thrust to drag relationship determined by aspect Ratio
    - ** The greater the number, the less drag, the faster the fish
    - ** Aspect ratio = (fin height)^2/fin area

       *   Rounded, truncate, forked, lunate, heterocercal

       *   Sharks use their fins to produce lift since they lack air bladders

  5.5.   Paired fins

       *   Bony fish use pectoral and pelvic fins for turning, balancing, and other fine maneuvers

       *   When swimming fast, these fins fold against body

  5.6.   Anal and dorsal fins

       *   Used for propulsion by a few fish

  5.7.   Propulsion by other nekton

       *   Shrimps and prawns use paired abdominal pleopods and tail fan

       *   Squids expel water through siphon

       *   Squids also use lateral fins like skates and rays

  5.8.   Speed

       *   *Stenella,* 40 km/hr (25 mi/hr)

       *   Killer whales, 55 km/hr

       *   Olympic swimmers, 5 km/hr

       *   Tuna, 74.6 km/hr (45 mi/hr)

       *   Red vs. white muscle

         **   Red, smaller cells rich in myoglobin, bind O_2 aerobic, metabolic rate 6X white muscle

         **   White muscle cells, larger, anaerobic, lactic acid

       *   Slow-moving fish (e.g., grouper) mostly white muscle

       *   Tuna, 50% red muscle fibers

  5.9.   Schooling

       *   Vary greatly in size, single species, similar in age and size

       *   Protection - broadcast spawners

6.0. Two Specialized Approaches to Feeding

  6.1.   Introduction

       *   Most nekton carnivores at second or higher trophic level

       *   Obtain large but rare prey or many small more common prey

  6.2.   White sharks

       *   Hunt pinnipeds or whales as adults

  6.3.   Baleen whales

       *   Planktonic crustaceans or shoaling fish

       *   Bowhead whales: fine, long baleen to collect *Calanus*

       *   Humpbacks: coarser baleen and feed on euphausiids, sand lances, capelin

       *   Gray whales: suck prey on bottom from one side of mouth and expel water out the other side

Journal Questions

1. Summary

Consider nekton in detail. Describe how nekton obtain oxygen, noting the differences between air-breathing and water-breathing animals. Note carefully how animals control buoyancy in the marine environment. Locomotion is critical to all nekton. Explore the role of body shape and fins to locomotion. Consider the role of schooling. Lastly, compare and contrast two approaches to feeding.

2. **Define the following key words in your own words.** Please feel free to add additional words to this list.

   a. countershading
   b. photophores
   c. apneustic breathing
   d. bradycardia
   e. septa of *Nautilus*

   f. physoclist
   g. rete mirabile
   h. frictional vs. form drag
   i. aspect ratio
   j. pleopods

3. **Question Section**

This is a long and complex chapter. You may have questions concerning how nekton obtain oxygen. Consider that the problems of marine mammals are very different from fishes. Swim bladders have a development and role that may be confusing. Do you understand why some fish fill their swim bladder with lipids while other fish get along without any swim bladder? The role of body shape and fins also may be confusing. Do you understand the concept of aspect ratio, and why fish with different ratios swim at different rates?

4. **Reflections**

Here we consider the problems faced by large sea creatures. Are the subjects of oxygen supply, buoyancy, and locomotion easier to understand or more interesting compared with the issues faced by the very small that we have previously addressed? What part(s) of this chapter do you find the most complex? How can you develop a strategy to ensure you learn all the material in this chapter?

Practice Questions

Multiple-Choice Questions (see back of book for answers)

1. Most epipelagic nekton are
   a. gentle, slow-moving vegetarians.
   b. photosynthetic autotrophs.
   c. carnivorous predators.
   d. each of the above are equally common.
   e. none of the above.

2. In marine mammals, breathing is different compared to terrestrial forms because marine mammals
   a. exhale and inhale much more rapidly.
   b. can remove much more oxygen from the air inhaled.
   c. have enhanced blood volume.
   d. all of the above.
   e. none of the above.

3. How is it that marine mammals that dive deeply do not suffer from nitrogen narcosis like people?
   a. Marine mammals' lungs collapse, and thus the air is not in contact with their alveoli.
   b. Marine mammals are not sensitive to nitrogen in the blood.
   c. Marine mammals pump excessive nitrogen out of their blood because of nitrogen pumps in their secondary lamellae.
   d. All of the above
   e. None of the above

4. An example of a rigid buoyancy tank is
   a. the pen of a squid.
   b. the chambers of a *Nautilus*.
   c. the physoclist swim bladder in a bony fish.
   d. all of the above.
   e. none of the above.

5. How does air enter the swim bladder in a typical adult marine fish?
   a. It enters through the pneumatic duct.
   b. It enters through the oval body.
   c. It enters through the esophagus.
   d. All of the above can occur.
   e. None of the above.

6. If a fish could reduce frictional drag alone to a minimum, its shape would be like a(n)
   a. pencil.
   b. ball.
   c. airplane.
   d. brick.
   e. cone.

7. Butterfly fish that live in reefs are tall and elliptical in cross section because this shape permits them to
   a. make fine-position adjustments as they move about the corals.
   b. accelerate quickly from predators.
   c. sustain high speeds when schooling.
   d. all of the above.
   e. none of the above.

8. Much of the push to move a typical bony fish forward comes from
   a. back and forth movements of the dorsal fin.
   b. oarlike motions of the pectoral fins.
   c. wavelike motions of the body wall muscles.
   d. all of the above contribute equally.
   e. none of the above.

9. When bony fish swim rapidly, pectoral and pelvic fins are
   a. at right angles to their body to help achieve lift.
   b. flapped like wings to achieve forward speed.
   c. folded against their bodies.
   d. all of the above.
   e. none of the above.

10. White sharks typically eat
   a. seals.
   b. copepods.
   c. small fish.
   d. birds.
   e. humans.

Complete the sentences below by writing the correct answer in the blank space (see back of book for answers).

1. Most epipelagic nekton show _____ coloration.
2. Some mesopelagic fish have large mouths and eyes and light-producing organs known as _____.
3. All marine vertebrates store and transport oxygen by means of the pigment _____.
4. Diving mammals can slow their heartbeat and shut down peripheral circulation. This condition is known as the _____ _____.
5. Fish increase their ability to remove oxygen from water by arranging blood flow in capillaries to be in the opposite direction to water flow, a _____ system.
6. A fish that is capable of swimming fast over a long distance would be expected to have a caudal fin with a(n) _____ (high, low, intermediate) aspect ratio.

Chapter 14   Nekton—Migration, Sensory Reception, and Reproduction

Chapter Outline

1.0. Introduction
- 1.1.   Some large, fast nekton migrate
- 1.2.   Integrates reproductive cycles with local and seasonal patterns of primary productivity

2.0. Migration
- 2.1.   Introduction
  - *   Distances oceanic
  - *   Increases range of resources; conditions for larval and juvenile stages may differ from conditions needed to sustain adult population
  - *   Patterns are often related to current direction
  - *   Known from tagging studies and attached transmitters
- 2.2.   Some examples of extensive oceanic migrations
  - *   Skipjack tuna: central Pacific spawning areas during summer; adolescent fish feed off the coast of North or South America
  - *   Sockeye salmon: anadromous; two years upstream until develop to smolts stage; migrate to ocean; mature over several years
  - *   Atlantic eel: catadromous
  - *   Green sea turtles lay eggs on Ascension Island; young travel to coast of Brazil
  - *   Gray whale migrates 11,000 miles round-trip each year
- 2.3.   Orientation
  - *   Biological clocks; day length may be a clue
  - *   Gray whales migrate near coast and use visual clues
  - *   Salmon may use olfactory clues
  - *   Fish may detect electric fields produced by ions moving through magnetic field

3.0. Sensory Reception
- 3.1.   Introduction
  - *   More than 5 senses—electroreceptive and magnetoreceptive
  - *   What about invertebrates?
- 3.2.   Chemoreception
  - *   Olfaction; some respond to a few parts per billion
- 3.3.   Vision
  - *   Marine vertebrates and cephalopods have camera eye
  - *   Differs from human eye chiefly in shape; must account for refractive power of water
    - **   Flattened in front
    - **   Lens that focuses by moving back and forth, not by changing shape
    - **   Cones vary, some have three types like us
- 3.4.   Equilibrium
  - *   Statocysts (invertebrates): fluid-filled structures containing statoliths
  - *   Two types of receptors: gravity and acceleration
  - *   Labyrinth organ in fish
    - **   Sac-shaped chambers = gravity detectors
    - **   Canals = acceleration detectors
- 3.5.   Sound reception
  - *   Sound spreads 5X faster in water than air
  - *   Fish: otoliths

* Lateral line system
3.6. Echolocation
* About 20% mammals orient by listening for reflected echoes
* Whales and pinnipeds use vocalizations for communication
* *Tursiops* uses pulses of clicks (800 per second) for ecolocating
* Low-frequency clicks to survey general surroundings
* High-frequency clicks for fine discriminations
* Small whales produce sounds with nasal plugs and air sacs associated with blowhole
  ** Melon concentrates sounds into beam; may stun fish
* Sperm whales: lower frequency and slower repetition rate
  ** More powerful, longer range
  ** Spermaceti organ - melon - 20% weight
  ** Frontal and distal air sacs and monkey's muzzle used
* Dolphins can hear 150,000$^+$ Hz, humans up to 20,000
  ** Middle ear encased in bony tympanic bulla
  ** Lower jaw sensitive to sound
3.7. Electroreception and magnetoreception
* Cartilaginous fish: tiny pores on head lead to ampulla of Lorenzini. Detect bioelectric fields
* May be present in tuna, salmon, some birds and whales

4.0. Reproduction
4.1. Nonseasonal reproduction
* Large whales breed every 2–3 years
* Grunion spawn 1st three hours after highest part of highest spring tides
4.2. From yolk sac to placenta
* Viviparity vs. oviparity
* Cod lays 15 million eggs per season
* Skates and rays produce only a few large eggs
* Some sharks and a few bony fish give birth to live young
* Ovoviviparity - male sea horses with brood pouches
* Sand tigersharks eat other eggs or developing siblings; in a year may reach 1 meter
4.3. Some alternatives to conventional sex ratios
* Guppies XX female, XY male, usually
* In fish, autosomal genes for sex determination also exist
* Salmon sex determination strongly affected by androgens or estrogens
* Because 1 male can fertilize many females, equal numbers of each sex results in "excess" males
* Sheepshead fish: sequential hermaphrodites = = > 5 females per male
* Tropical cleaner fish of Great Barrier Reef occurs in social groups of 10 fish with only 1 functional male. The most aggressive = polygyny. If male dies, most dominant female transforms into a male
4.4. Reproduction in marine mammals
* Relatively small newborns face problem of heat loss
  ** Pinnipeds may keep pups on land for a while
  ** Gray whales give birth in warm waters
* Weddell seal and elephant seal pups double their weight in 2 weeks
* Blue whales: 3 tons birth --> 23 tons in 7 months
* Cetacean milk is 25–50% fat; produce 600 liters/day (baleen)
* Gestation period is 3–12 months; mate every 1–3 years
* Migratory species give birth and mate in warm waters. Cold water feeding grounds
* Northern fur seals: estrous few days after birth; delayed implantation; dormant blastocyst 4 months; placental connection; 7 months gestation

* Sexual dimorphism: elephant seals, fur seals, sea lions. Polygynous social order: most aggressive and vigorous males breed, hence, male makes 25X genetic contribution
* Genetic variation in many pinniped populations damaged by hunting

Journal Questions

1. Summary

Review in detail the migration of large nekton. Consider how sensory reception functions in marine animals. Note how sensory reception in the marine world is similar and/or different from terrestrial environments. Consider the variety of reproduction methods in nekton. How is reproduction related to migration?

2. **Define the following key words in your own words.** Please feel free to add additional words to this list.

a.   smolt
b.   catadromous
c.   elver
d.   *Chelonia*
e.   statolith
f.   labyrinth organ

g.   nasal plugs vs. monkey's muzzle
h.   spermaceti organ
i.   tympanic bulla
j.   ampulla of Lorenzini
k.   oviparity

3. **Question Section**

Are the biological advantages of migration clear to you? Are the different examples of migrations understandable? The details of sensory reception may not be completely understood. Do you have questions concerning reproduction of large nekton?

4. Reflections

This is an interesting and complex chapter. Are learning examples of migratory life cycles similar or different from learning details of sensory reception? Are these two topics linked? Again, we focus on the biology of large animals. Does that make this material more easy or more difficult to learn than topics related to the biology of very small organisms?

Practice Questions

Multiple-Choice Questions (see back of book for answers)

1. Given the reasons stated in your text concerning the value of oceanic migrations, which statement below is most likely to be true?
    a. Adult whales that migrate may not eat at the site where they bear their young.
    b. Adult whales are likely to eat most of their food at the same location where they bear their young.
    c. The migration pattern of fishes is usually opposite to the oceanic currents.
    d. Pelagic animals that migrate usually hug the coastlines.
    e. Our knowledge of migration routes is largely known from direct observations throughout the journey of the animal in question.

2. Where do catadromous Atlantic eels hatch?
    a. in freshwater streams
    b. at sea
    c. in brood pouches found in the male fish
    d. all of the above
    e. none of the above

3. Why do many whales bear their young in warm waters?
    a. Subtropical and tropical waters have more small fish upon which young whales feed.
    b. Subtropical and tropical waters have more available food for adult whales.
    c. Young whales are relatively small and thus may lose too much heat in cold waters.
    d. all of the above
    e. none of the above

4. Which migration is likely to involve an animal following the landmarks of the shoreline?
   a. adult green turtles south of the Sargasso Sea
   b. Atlantic eel
   c. Pacific skipjack tuna
   d. all of the above
   e. none of the above

5. Fish may be able to detect disturbances in the surrounding water caused by prey by means of
   a. labyrinth organ.
   b. rods.
   c. statocysts.
   d. lateral line system.
   e. taste buds.

6. Why do sperm whales produce different sounds for echolocation compared with smaller toothed whales?
   a. Sperm whales feed on prey that live much deeper.
   b. Sperm whales communicate by a different language.
   c. Sperm whales have poor eyesight.
   d. all of the above
   e. none of the above

7. Animals that lay eggs are said to demonstrate
   a. viviparity.
   b. oviparity.
   c. polygyny.
   d. androgeny.
   e. ontogeny.

Complete the sentences below by writing the correct answer in the blank space (see back of book for answers).

1. Salmon born in streams and spending much of their adult lives at sea are said to be _____.
2. Fishes that live in deep water may exclusively use cells called _____ to detect light.
3. Fluid-filled organs that respond to gravity are called _____.
4. Vertebrates know their rate of speed by means of the _____ organ.
5. Clicks produced by whales are focused by means of a fatty lens-shaped _____.
6. The external auditory canal may not play a significant role in hearing by whales; instead, they seem to use their _____ _____.
7. The time between fertilization and birth is the _____ period.

Chapter Outline

1.0. Food from the Sea
     1.1.   Human population reached 1 billion in 1850
     1.2.   Now 6^+ billion; 0.5-1.0 billion undernourished
     1.3.   Demands for all resources enormous
     1.4.   Hence, the sea is heavily exploited for food, recreation, military purposes, shipping, extraction of oil and gas, and other minerals—an international problem
     1.5.   In such a world, marine resources are critical
     1.6.   1990, world fishing = almost 40 billion-dollar industry
     1.7.   Relatively few species have been focused on, hence, many of these species have been eliminated from traditional fishing ground or face extinction

2.0. A Brief Survey of Marine Food Species
     2.1.   Clupeoid fish = anchovies, herrings, sardines, etc., = 1/3 of total commercial catch
         *   Peruvian anchoveta = 19% of 1970 catch; overfishing caused fishery to collapse a few years later
         *   Most fish meal used to supplement livestock and poultry fodder
         *   Schooling behavior makes them easy to catch with purse seines (600 m long and 200 m deep); trap entire school
     2.2.   Gadoid fish = cod, pollack, hake; 12^+ million tons
     2.3.   Tuna: top of food chain; 7 or more trophic levels
         *   Yellowfin tuna associated with dolphins
         *   1972, 300,000 dolphins dead by United States fleet. Then, Marine Mammal Protection Act; reduced to 20,000 dead. Use longline methods

3.0. Major Fishing Areas of the World Ocean
     3.1.   Bottom fish (halibut, flounder, sole) + benthic invertebrates = 15% of total catch in shallow near-shore waters
         *   High prices and stable markets result in extensive fishing for these animals
         *   North Atlantic overfished; 90% from waters over continental shelf
         *   Antarctic fishery still not exploited; lack of nearby population and great distances to processing facilities

4.0. A Perspective on Sources of Seafoods
     4.1.   Human diet mostly cereals, vegetables, fruits, beef, poultry, pork, and fish
     4.2.   Most seafoods are animals 3-4 trophic levels above primary produces, terrestrial planting avoids this loss of energy
     4.3.   Seafoods are wild, unimproved stocks, hunted rather than controlled and domesticated. Little has been done to improve fish stocks
     4.4.   1955, 86% world catch eaten directly and 14% fish meal for animals
     4.5.   1990, 40% fish meal—only 20% of this edible pork and poultry
         *   Thus, not 83 million tons of food for people but 61 million
     4.6.   Long food chains, unsophisticated production systems, etc., = = > marine environment does not produce much food = 1% of food eaten by people

4.7.   From W.W.II --> 1970, global fish catch increased 6% per year while human population increased 2% per year
* 1970 --> present, gains reduced to 1.8%
* Are we approaching a limit to growth?
* Magnitude of harvest limited by several factors
** Number of trophic levels in food chain leading to fish in question
** Efficiency from one level to another
** Fish must be of a minimal size for commercial harvest

4.8.   In open ocean, 3-4 trophic levels to produce fish a few cm in length

4.9.   Tuna or squid = 5 trophic levels

4.10.   Average trophic levels = 3 for coastal waters and 1.5 for upwellings areas

4.11.   For sustained yield, 1/2 to 2/3 of fish stock must escape

4.12.   Upshot, with current methods, fish as a resource may provide 2-3X present yield—not much multiplied by 1% of human diet, especially considering increases in human population

4.13.   Mariculture
* Application of farming techniques to grow, manage, and harvest marine animals and plants
* Simple enhancement—raise and release juvenile fishes
* Intensive captive maintenance for entire life span
* In the East, mullets and milkfish are raised in shallow estuarine ponds
* Salmon ranching in Japan, Norway, and United States, raised and released 3 billion smolts; 1-2% return = 30 million salmon. More difficult for other species. How can the farmer be assured that she/he will get the return?

4.14.   Moving down the Food Chain
* Oysters, mussels, abalones, lobsters, and red algae grown under controlled conditions
* Expensive—restricted to luxury items. Adds nothing to the diets of those who need it most
* In theory, if we could harvest what fish eat, food production could increase by a factor of ten
* However, the technology to harvest plankton would require more energy, and in the end, it may not be worth the effort
* Once, the blue, fin, and humpback whales around the Antarctic were a way to harvest krill. The Japanese and former Soviets estimate that one may take 100-200 million tons of krill from these waters. What do you do with krill?
* Upshot, for the immediate future, the growing human population will depend primarily on terrestrial food production

5.0. The Problems of Overexploitation

5.1.   When a new stock identified, first catches are large and contain many large fish

5.2.   Then, average size and numbers decrease

5.3.   If possible, the maximum sustainable yield should be taken

5.4.   In practice, many stock overfished

5.5.   The Peruvian anchoveta
* Incas began first commercial exploitation - indirect - guano deposits from nesting seabirds. Fertilizer for subsistence farmers
* 1950, fishing began; 1951, 7,000 tons landed; 1970, 13 million tons = 20% total world catch
* Response = dramatic drop in seabird and fish populations

5.6.   The great whales
* Aboriginal hunting has occurred for thousands of years
* 18th and 19th centuries, whale oil profitable
* End of 19th century, gray, right, and bowhead on verge of extinction
* IWC set quotas after W.W.II, but they were never met
* Fate of great whale & Peruvian anchoveta remain unknown

* 1982, IWC declared moratorium for 5 years beginning in 1985. Opposed by Japan, Norway, and Iceland
* 1991, IWC members opposed to indefinite moratorium began limited whaling of minke and fin whales

6.0. The Tragedy of Open Access
    6.1. Hardin's ''tragedy of the commons:'' rewards most those who exploit the most

7.0. International Regulation of Fisheries
    7.1. A variety of international commissions have failed so far to avoid overfishing of cod and other fish
    7.2. 1976, United States and the Exclusive Economic Zone, 200 mile limit
    7.3. 1982, Law of the Sea
        * 200-mile EEZs
        * No mention of Antarctic krill; last unspoiled waters on earth; first effort to recognize integrity of large marine ecosystem

Journal Questions

1. Summary

Review in detail the major marine food species and the major fishing areas. Consider the role of the sea as a present and future source of food for people. Consider the issues that led to overexploitation of marine resources and international efforts to avoid the ''tragedy of the commons.''

2. **Define the following key words in your own words.** Please feel free to add additional words to this list.

a. clupeoid vs. gadoid fish
b. mariculture
c. krill
d. maximum sustainable yield

e. International Whaling Commission
f. open access
g. Law of the Sea

3. **Question Section**

Do you understand why marine resources provide us with only 1% of our food, even though primary production in the sea could, in principle, provide for much more of human needs? Are you clear on the relationship between the length of a food chain and how efficient it is for food production? Do you appreciate the relationship between the "tragedy of the commons" and the need for international regulation of the fishing industry?

4. Reflections

In this chapter we consider the practical value of the sea considering certain ecological concepts. Does the relationship of fisheries to a hungry world make this subject easy to relate to and learn? Did any sections of this chapter surprise you? Why? What challenges are presented to your generation?

Practice Questions

Multiple-Choice Questions (see back of book for answers)

1. Why is the pressure on the sea as a source of food likely to increase?
    a.   At least 0.5 billion people are currently undernourished.
    b.   The current population of the earth places far greater demands on resources compared with far smaller previous populations.
    c.   There is a trend to use marine resources to feed land-based farm animals.
    d.   All of the above contribute to the problem.
    e.   None of the above, because of newer methods of grain production, the need for ocean resources is likely to diminish.

2. Clupeoid fish are one-third of the world catch and are mostly used for
    a.   frozen fish sticks and other similar products eaten by people worldwide.
    b.   livestock and poultry fodder.
    c.   feeding the poor people of the Third World.
    d.   all of the above.
    e.   none of the above.

3. Why is tuna fishing considered a problem?
    a.   Fishing for yellowfin tuna by purse seiners has resulted in the deaths of many dolphins.
    b.   Tuna feed directly on phytoplankton, and if the tuna are removed from the sea, phytoplankton populations may bloom and form toxic zones.
    c.   Tuna accumulate large amounts of toxic waste and are not fit for human consumption.
    d.   all of the above
    e.   none of the above

4. Compare the number of steps in the food chains to create a tuna and a cow.
   a. The number of steps in the food change is the same.
   b. A cow requires at least four steps, and because tuna eat phytoplankton, only two steps are at issue.
   c. A cow that eats grass embraces two trophic levels, while tuna that eat small zooplankton may include three thropic levels.
   d. All of the above may be true at different times.
   e. none of the above

Complete the sentences below by writing the correct answer in the blank space (see back of book for answers).

1. In 1970, 19% of all fish caught belonged to a single species, the _____ _____.
2. A type of benthic fish, caught in shallow waters, that commands a high price is _____.
3. Given the present production methods, about how much of a fish stock must be permitted to escape in order to continue to harvest them indefinitely? _____
4. The organization that oversees the utilization and conservation of pelagic whales is the _____ _____.
5. The notion that a coastal state has an exclusive jurisdiction to all fish within 200 miles of its coast was a feature of what United Nations Conference? _____ _____ _____

Chapter **16**   Ocean Pollution

Chapter Outline

1.0. Introduction
    1.1.   Growing human population stresses ocean with domestic and industrial waste
    1.2.   Much dumped into semienclosed estuaries, bays, or lagoons
    1.3.   Chemicals used to produce food - pesticides and fertilizers among the greatest polluters
    1.4.   The effects of contaminants, such as heavy metals, radioactive waste, and pesticides, are magnified as they are transferred up the food chain

2.0. Sewage
    2.1.   Fouls beaches and bays with toxic and nontoxic pollutants
    2.2.   Despite sewage treatment plants, the growing coastal communities are increasing discharge into oceans and estuaries
    2.3.   In the United States, 1,300 major industries and 600 municipal wastewater treatment plants
        *   45 million tons of sewage sludge discharged into coastal marine environment
    2.4.   In undeveloped world, discharge of raw sewage and sludge into estuaries and ocean is common
    2.5.   In northeast United States, sludge dumped at sea, New York City dumps 160 km off New Jersey. Sludge contaminated with heavy metal and human bacteria; effects under study
    2.6.   Kaneohe Bay in Hawaii: 1950–1975, 2/3 of coral reefs damaged by green algae that grew in response to nutrients in sewage
    2.7.   Southern California: 1/4 million tons of sludge per day dumped at 30 outfalls
        *   DDT and PCBs increased several fold in sediments and benthic invertebrates
        *   Habitats changed or degraded

3.0. Toxic Pollutants
    3.1.   Introduction
        *   Toxins enter sea through sewer systems and industrial discharges
        *   Effects of many unknown
    3.2.   Antifouling paints
        *   Boat hulls sheathed in copper for centuries to retard attachment of barnacles
        *   Recently, tributyltin (TBT) found to be more effective
            **   Leaches into harbors and is undetectable with current technology
            **   A few parts per billion can kill shellfish
            **   1987, restrictions on use
    3.3.   DDT
        *   Dichloro-diphenyl-trichloroethane
        *   1945, first of new class of synthetic chlorinate hydrocarbons
        *   Kills houseflies, lice, mosquitoes, and crop pests
        *   Benefactor --> ecological nightmare. Banned in United States and other countries
        *   Absorbed by particles, nearly insoluble in water
        *   Worldwide problem; even in tissues of penguins of Antarctic
        *   Fat soluble, enters food chain at level of phytoplankton. Bioaccumulation eventually reaches top carnivores
        *   1960s, Los Angeles outfall, 100 tons per year
        *   Birds, 4–5 trophic levels removed from phytoplankton, may suffer the worst. Kills directly or prevents calcium deposition in egg shells. Pelican population reduced in California
        *   United States banned use in 1972, but still used in other parts of the world

3.4. Dioxins
* Another chlorinated compound; toxic to birds at a few parts per quadrillion
* Most arise from manufacturing of paper
* Bioaccumulation in fatty tissues of fish; may cause problems in humans: cancers, malformations, immune system and reproductive difficulties
3.5. PCBs
* May suppress the immune system; linked to 1988 seal kill in North Sea
* Also linked to implantation problems and lower birth-weight pups

4.0. Oil on Water
4.1. Oil spills, such as *Exxon Valdez* in 1989, suffocate benthic organisms and kill birds by reducing insulating value of feathers
* *Valdez* on Bligh Reef in Alaska's Prince William Sound; 242,000 barrels = 40,000,000 liters. Over 26,000 square km
* 33,000 dead birds recovered
* However, we need to reduce dependency on foreign oil.
4.2. More oil enters as runoff from roads, leaking underground storage tanks, dumped waste oil, and bilge water

5.0. Marine Debris
5.1. Introduction
* Before 1987, foreign and domestic merchant ships, military and fishing boats, and recreation boats dumped massive amounts of garbage at sea, including plastic
* 1987, Marine Plastics Pollution Research and Control Act (MARPOL), directs United States EPA to find ways to prevent plastics pollution
* However, ocean dumping seems necessary, and plastic floats; medical wastes are also becoming a problem
5.2. Fishing gear
* Fishing boats lose 100,000 tons of nets and other gear each year
* From 1940, natural fibers --> plastic, which is nonbiodegradable
* Birds, seals, and others entangle and die
5.3. Plastics
* Jugs, bottles, buckets, bags, sheeting, eating utensils, yokes of 6-packs, life preservers, buoys, fish nets, styrofoam cups and packing material, suspension beads—beaches covered by plastic waste

6.0. Concluding Thoughts: Developing a Sense of Stewardship
6.1. In the next 15 years, human population on coastal areas will increase 50%
6.2. We position ourselves on the top of heavily exploited marine food chains while we contaminate them with materials too dangerous to place near our homes
6.3. . . . harmonize our civilization with the environment so that our children see our wisdom, not inherit our wastes.

Momiji

Journal Questions

1. Summary

Review the major agents of ocean pollution.

2. Define the following key words in your own words. Please feel free to add additional words to this list.

a. primary vs. secondary treatment of sewage

b. tributyltin

c. DDT and bioaccumulation

d. dioxins

e. PCBs

f. *Exxon Valdez*

3. Question Section

Do you understand the chemistry of the various agents described in this chapter, and how they affect the marine environment? Do you understand the relationships among development, pollution, and stewardship?

4. Reflections

Are you more motivated to understand aspects of chemistry when the health of the marine environment is at issue? In general, does writing in this journal help you to learn? Have this chapter and the rest of this course affected your personal notion of stewardship?

Practice Questions

Multiple-Choice Questions (see back of book for answers)

1. DDT
   a. is a copper-based antifouling paint.
   b. is a synthetic chlorinated hydrocarbon originally used to kill insect pests.
   c. spilled out of the *Exxon Valdez* and killed large numbers of animals off the coast of Alaska.
   d. all of the above.
   e. none of the above.

2. Why are plastics a threat to marine life?
   a. Plastic nets that are lost float indefinitely and entangle many animals.
   b. Animals that swallow plastic bags and small plastic pieces may be injured.
   c. Six-pack yokes may get onto the necks of seabirds.
   d. all of the above
   e. none of the above

Complete the sentences below by writing the correct answer in the blank space (see back of book for answers).

1. Solid waste that forms during primary and secondary treatment of sewage is called _____.
2. Toxic agents that suppress the immune response of sea animals include _____ and _____.

Answer Key

Chapter 1

Multiple Choice

1. d
2. c
3. a
4. c
5. c
6. e
7. d
8. b
9. a
10. d
11. b
12. a
13. b
14. a
15. b

Fill in the Blank

1. neritic
2. aphotic
3. upwelling
4. countercurrent
5. neap
6. nitrate, phosphate
7. wind
8. buffer
9. continental slope
10. ultraviolet

Chapter 2

Multiple Choice

1. a
2. e
3. a
4. c
5. d
6. a
7. a
8. d
9. c
10. d

Fill in the Blank

1. chloroplasts
2. ecological
3. asexual
4. isomotic
5. excrete
6. diffusion
7. poikilotherms
8. anaerobic
9. detritus
10. suspension feeders

Chapter 3

Multiple Choice

1. c
2. d
3. e
4. d
5. a
6. c
7. c
8. a
9. e

Fill in the Blank

1. shorter wavelengths
2. phycobilins
3. Coccolithophores
4. epitheca, hypotheca
5. Dinophyta

Chapter 4

Multiple Choice

1. b
2. a
3. c
4. b

5. a
6. c
7. a
8. b

Fill in the Blank

1. holdfast
2. Phaeophyta, Rhodophyta
3. Rhodophyta
4. xanthophylls
5. sporophyte
6. meristematic
7. rhizomes
8. warm, cold

Chapter 5

Multiple Choice

1. e
2. d
3. b
4. d
5. d
6. c
7. b
8. c
9. e

Fill in the Blank

1. pelagic phytoplankton
2. turnover rate
3. chlorophyll *a*
4. photoinhibition
5. ATP, $NADPH_2$
6. accessory
7. calcium carbonate
8. upwelling

Chapter 6

Multiple Choice

1. c
2. d
3. b
4. c
5. a
6. c

7. a
8. a
9. b
10. e
11. d

Fill in the Blank

1. water vascular system
2. pentamerous
3. horseshoe crab
4. Annelida
5. cephalization
6. Nematoda
7. bilateral
8. ctenes
9. polyp
10. spicules

Chapter 7

Multiple Choice

1. a
2. d
3. c
4. b
5. b
6. a
7. b
8. c
9. d
10. a

Fill in the Blank

1. hollow, dorsal
2. fishes
3. anadromous
4. urea
5. Amphibia

Chapter 8

Multiple Choice

1. e
2. a
3. d
4. a
5. b

6. a
7. d
8. b
9. b
10. d

Fill in the Blank

1. mangroves
2. grazing, detritus
3. cyanobacteria
4. *Littorina*
5. mussels, or *Mytilus*
6. biological succession
7. H_2S
8. circadian rhythm

Chapter 9

Multiple Choice

1. d
2. b
3. d
4. a
5. a

Fill in the Blank

1. flushing time
2. osmotic conformers
3. mangrove

Chapter 10

Multiple Choice

1. a
2. d
3. e
4. a
5. d
6. c
7. d
8. a

Fill in the Blank

1. fringing
2. lagoon
3. broadcast spawners

4. planula
5. light
6. giant clam, or *Tridacna*
7. cleaning

Chapter 11

Multiple Choice

1. c
2. d
3. b
4. a
5. c
6. d
7. d

Fill in the Blank

1. brown algae, or kelps
2. a similar number of
3. cropper
4. H_2S
5. hemoglobin

Chapter 12

Multiple Choice

1. a
2. d
3. b
4. e
5. b

Fill in the Blank

1. neuston
2. vertical migration
3. deep scattering layers
4. mucus

Chapter 13

Multiple Choice

1. c
2. d
3. a
4. b
5. e

6. b
7. a
8. c
9. c
10. a

Fill in the Blank

1. countershading
2. photophores
3. hemoglobin
4. diving reflex
5. countercurrent
6. high

Chapter 14

Multiple Choice

1. a
2. b
3. c
4. e
5. d
6. a
7. b

Fill in the Blank

1. anadromous
2. rods
3. statocysts
4. labyrinth
5. melon
6. lower jaw
7. gestation

Chapter 15

Multiple Choice

1. d
2. b
3. a
4. e

Fill in the Blank

1. Peruvian anchoveta
2. halibut, flounder, or sole
3. 1/2 to 2/3
4. International Whaling Commission
5. Law of the Sea

Chapter 16

Multiple Choice

1. b
2. d

Fill in the Blank

1. sludge
2. PCBs and dioxins

About the Author

Dr. Larry Lewis began his interest in marine biology while an undergraduate student at Southampton College. He received graduate training in marine botany at the University of South Florida and the New York Botanical Garden before he received a Ph.D. in Biology from Fordham University. Subsequently, Dr. Lewis spent five summers at the Marine Biological Laboratory at Woods Hole and once served as Director of Education for the Ocean Research and Education Society (ORES). ORES gave college students an opportunity to study humpback whales in the North Atlantic from the deck of a tall ship, the *Regina Maris*.

Currently, Dr. Lewis is an Associate Professor of Biology at Salem State College and president of the Executive Board of the Massachusetts Bay Marine Studies Consortium—a consortium of 20 institutions in the Boston area. He teaches marine biology each summer at the University of Massachusetts at Lowell.